hamlyn

Guide to

Wild Flowers

of Britain
and Europe

D. & R. Aichele
H. W. & A. Schwegler

e

ıry

Acknowledgements
The publishers wish to thank the following for their permission to reproduce their photographs: D. Aichele, R. Aichele, J. Apel, F.G. Averdieck, C. and W. Baitinger, H. Bechtel, W. Bechtel, F. Büttner, N. Caspers, R. Fieselmann, K. Frantz, E. Garnweidner, M. Haberer, H. Haeupler, D. Herbel, E. Humperdinck, F. Jantzen, W. Käckenmeister, R. König, P. Kohlhaupt, Kullmann, H.E. Laux, Ch. Lederer, E. Müller, G. Quedens, G.H. Radek, G. Rein, H. Reinhard, S. Sammer, W. Schacht, J. Schimmitat, H. Schmidt, T. Schneiders, P. Schönfelder, H. Schrempp, F. Schwäble, F. Siedel, M. Stellwag, K.F. Wolfstetter and P. Zeiniger.

The publishers are grateful to Dr Robert Press, Kenneth Beckett and Gillian Beckett for their assistance.

Foreword and Introduction acknowledgements
S & O Mathews 11 Top, Natural Image/Bob Gibbons 11 Bottom, 17 Bottom, 19, N.H.P.A./Brian Hawkes 13/E A Janes 17 Top/Dr Eckart Pott 2, Jason Smalley 4, 15 Top, 15 Bottom

This revised edition first published in Great Britain in 2001 by
Hamlyn, an imprint of Octopus Publishing Group Ltd
2–4 Heron Quays, London E14 4JP

First published in 1992
by The Hamlyn Publishing Group Limited
Reprinted 1993

Original title: *Was grünt und blüht in der Natur?*
Translated by Susan Kunze

ISBN 0 600 60243 5

A CIP catalogue record for this book
is available from the British Library

Design @ 2wo
Printed in Italy
by Rotolito Lombarda

Contents

Foreword **5**
Introduction **6**
Key to abbreviations and symbols **9**
Habitats **10**
    Weedy and uncultivated areas **10**
    Grasslands **12**
    Wooded areas **14**
    Wetlands **16**
    Pioneer communities **18**
Identifying plants **20**
Pictorial key and plant identification **23**

Plants with white flowers 32
Maximum four petals **32**
Five petals **48**
More than five petals **78**
Bilaterally symmetrical flowers **94**
Tree or shrub **104**

Plants with yellow flowers 118
Maximum four petals **118**
Five petals **132**
More than five petals **156**
Bilaterally symmetrical flowers **178**
Tree or shrub **198**

Plants with red flowers 206
Maximum four petals **206**
Five petals **214**
More than five petals **238**
Bilaterally symmetrical flowers **252**
Tree or shrub **282**

Plants with blue flowers 292
Maximum four petals **292**
Five petals **304**
More than five petals **322**
Bilaterally symmetrical flowers **330**

Plants with brown or green flowers 350
Maximum four petals **350**
Five petals **366**
More than five petals **370**
Bilaterally symmetrical flowers **372**
Tree or shrub **378**

Glossary **382**
Index **385**

Foreword

Concern for our environment is not just a modern trend, it is vitally important. We must, of course, be able to offer our children and our children's children secure jobs and healthy places to live, but sufficient countryside and other natural landscapes in which to pursue leisure-time activities are also essential.

When we talk about the environment, most of us think of the world of plants: woods, heaths, moors and meadows; valleys, hills and mountains; streams, rivers, lakes and the seashore. When we think of these places we think of the plants, trees and shrubs that grow there. Whenever we try to re-create nature in our cities, parks and greenbelts are planned, landscaped and stocked with an abundance of flowering and non-flowering species. A modern idea in gardening is to plant an area with 'wild' grasses and flowers – an attempt to bring the countryside into our backyards.

Plants, trees and shrubs tell scientists a great deal about the areas in which they can be found. The primrose, for example, does not grow well in soil with lime in it, and so we know that where this plant can be found growing the soil will not contain this element. This is a simple example of how a plant can give us accurate information about its natural habitat. When taken as a whole, the plants in any given area will tell a very revealing story about their landscape. It is not surprising, therefore, that an increasing number of people want to know more about plants and want to be able to identify them when they see them.

The most obviously memorable part of a plant is usually its flower, and this book is arranged so that you can easily look up the flower colour using the coloured band on the edge of the pages. The flower form (that is, the number of petals) then limits the number of possible species within each colour. You should then look through the photographs of the plants that fall into the same categories of colour and form to find the specimen that you wish to identify. Having found it, the symbols and text that accompany each photograph either confirm the identification or direct you to similar species for a correct identification. The text gives a great deal of information on important identification marks, habitat and distribution.

Although this system of identification might be considered unscientific, it has helped a great many amateur naturalists to identify species that they have wanted to learn more about and to increase their knowledge of the natural world. With this system it has been possible to keep the book to a handy size, while at the same time including well over 600 species each with full-colour photographs (only a few species are duplicated because of their differing flower shades). There are also references to similar species so that almost 1,000 types are covered, and this far exceeds the usual number of plants described in other popular books on flora.

Amateur botanists nowadays travel extensively throughout Continental Europe, and for this reason certain interesting plants have been included which may be found only in specific places on the Continent. Similarly, plants which may grow wild in one country, some of the mountain flowers for example, may be well-known cultivated rockery plants elsewhere. Quite frequently, cultivated flowers 'escape' and become established in the wild, hence their inclusion in this book.

Also featured are numerous flowering shrubs and trees, some truly wild, others more generally cultivated. Finally it should be mentioned that in some countries certain species are protected due to their rarity, whereas elsewhere they may flourish profusely and not need to be protected. This important point is covered in the Introduction where the system of symbols accompanying each plant description is explained.

Introduction

The plant descriptions in this book have been arranged in such a way that they form a unit with the photographs directly above them. This simplifies the process of identifying any species.

Identification marks

This section contains a description of the main features of each specimen. Easily recognizable characteristics that distinguish a plant from any other similar species are given. This information is intended to complement the photographs.

Habitat and distribution

The second section gives valuable data on the plants' soil and climate requirements. Details of the types of landscape where each species is normally found are also given, as well as the type of vegetation in which the plant is likely to grow.

This section also tells the reader how often he or she might come across the specimen concerned. A plant is referred to as *common* when it can be found in many different places within its area of distribution, as *local* when it appears only to be found here and there, and as *rare* when it can only be found in very few places. No reference is made to the concentration in which the plant can be found in its habitat. It may well be that even rare plants can be found in abundance in one particular area or even only in one individual spot.

Additional information

The last section gives distinguishing details of related species or even of particular subspecies, plus general points of interest. Synonyms additional popular names plus scientific names are also included where appropriate.

Scientific names

The problem of the correct scientific name is an old one which even today has not been settled satisfactorily. In the mid-eighteenth century the Swedish naturalist Carl von Linné introduced double names for each species: the first name indicates the genus, the second is the name of the species (which is always written with a small initial letter). Often the second part is the name of the person who first identified the species, and after whom the plant has been named. This method should have made it easy to name any plant unmistakably and clearly. Experience proved that this was not so, however. Often the same species was 'discovered' for the first time by more than one person and named differently by each one. Other species were subdivided or amalgamated using different methods. Some scientists objected to a name on linguistic grounds and quickly created another. Others felt that the old name was not descriptive enough or that it was incorrect. The net result was that many plants had a whole range of names.

Only through the strict application of rules which were worked out laboriously in a series of international congresses has it been possible over the last few decades to bring relative (but not yet final) stability to the situation. The plants in this book are named according to *Flora Europaea*, volumes 1–5 (Cambridge University Press), the most up-to-date single work regarded as authoritative for the whole of Europe. This may mean that some well-known and popular names have not been included, but it is a step towards final standardization.

Common names

The situation with common names is different because these are not governed by any rules. As can be seen from other books on plants, names vary from area to area. The names given here are those recommended by the Botanical Society of the British Isles in *English Names of Wild Flowers*, Dony, Robb and Perring (Butterworth). Whenever a plant was not included in this work, names from other books on wild flowers were used. While the scope of this book is mainly British, it is wide-ranging throughout Europe, and a few species have no common English name.

Family name

The name of the family to which a species belongs is given below the species name itself. Difficulties are far less frequent here. In some cases the old name is given in brackets where this is well known.

Further information

Below the various plant names additional information and symbols are given which are easy to understand and which provide further aids to identification.

Flowering time

The majority of individual plants of one species are found to be in blossom during a specific period, and the months at the beginning and end of the usual flowering time for each species are quoted. Flowering time does vary somewhat depending on annual climate and geographical position including height above sea level, but within certain tolerances this period is relatively firmly established. It is not usual to find snowdrops, for example, growing wild in bloom in summer. Individual early or late blossoms are exceptions which do occur occasionally. However, a spring plant may have a late blossoming. With some of these species the start of blossoming depends on the length of the day and in spring this is very short (short-day plants). If the autumn is mild and the days as short as at certain times in spring, then there may well be a widespread second flowering in one year.

Height of growth

The figures given below the flowering time indicate to what height a plant will grow under normal conditions. Plants tend to grow to a certain 'level', depending on the type of plant and prevailing conditions. In a meadow, for example, some species remain very close to the ground and even when in flower they are hardly more than 20 cm high. Others grow to the level of the lower grasses, about 50 cm high. The remainder will compete for space and light with the tall grasses.

The height of a plant is a distinguishing feature and should not be ignored for the purposes of identification. In time, the amateur naturalist gets to know whether the conditions in which he or she finds a plant are its normal habitat, or whether a plant has found itself in a spot which causes it to produce abnormal growth. A lack of light or an excess of nourishment (overfeeding) can lead to abnormally tall examples of species. Where the soil is poor or the plant is in an exposed position, growth may be less than usual, but nevertheless the plant can be quite sturdy. This should not deter anyone from making a positive identification. Compare the conditions in which a plant is found to be growing with those in which it is normally found and an explanation for excessive or reduced growth will often become evident.

Classification

Genuine flowering plants (the angiosperms) are divided into two subclasses. Monocotyledons are plants whose embryos have one cotyledon. The more extensive subclass, Dicotyledons, have two cotyledons in the embryos. These are further divided into three groups: those with flowers with fused petals (gamopetalous), those few flowers without petals (perianth absent), and the large number that have free petals (polypetalous). Gamopetalous species display petals which have to a greater or lesser extent grown together to form a mass.

The life cycle of plants

Different groups of plants have their own life cycle, that is from germination, through the growing, flowering and fruiting periods, to death.

ANNUAL PLANTS are those which germinate, flower once, fruit and die completely within one year.

BIENNIALS produce only leaves in the first year (storage) and blossom, fruit and die in the second.

Plants which live and blossom repeatedly for several years are divided into two groups:

HERBACEOUS PERENNIALS 'die back' each autumn, after flowering and fruiting, which means that the part of the plant that can be seen above the ground withers and dies, but below ground the rhizomes, tubers, bulbs, etc. remain and produce new shoots each year.

WOODY PLANTS, such as trees and shrubs, produce woody parts above the ground which remain and grow taller in each vegetation cycle. They either retain their leaves throughout the year (evergreen) or produce new leaves each spring, which wither and fall in the autumn (deciduous).

Poisonous plants
Resistance to poison varies from person to person. With some plants contact alone causes allergic rashes (stinging nettles, types of primroses). Even chewing or eating harmless plants can cause ill-effects.

Flower colour and form
The coloured band at the top of each page indicates the colour of the flowers in each section. These sections have been further divided according to the flower form. Plants with radially symmetrical flowers, such as daisies, are described first and have been further subdivided:

 Flowers with up to 4 petals

 Flowers with more than 5 petals

 Flowers with 5 petals

 Plants with bilaterally symmetrical flowers, such as violets and Lady's-slipper, follow on from the radials.

 This indicates flowering trees or shrubs and it precedes the symbol for the flower type itself, to make identification quicker.

A complete and simplified key appears on pages 20–21.

Protected species
Species which ought to be protected in some way are indicated by either a green or a black triangle (the latter indicating a species protected by law in Britain, but not necessarily protected in Europe). Protection of plants and animals ensures that we leave our environment intact for future generations. It is our responsibility therefore to treat the natural world around us with respect. Plants given the triangle symbol may or may not be protected by law in the countries in which we find them, but they have been noted to be in decline by botanists or other scientists and therefore need preserving.

Habitat symbols
The symbols for habitats are wider ranging than for Britain only and have been allocated on a northern European basis. Where British habitats differ from those indicated by the symbols, you will find this information in the Habitat and distribution section of the text.

Key to abbreviations and symbols

⊞	Radially symmetrical flowers with up to 4 petals
⊞	Radially symmetrical flowers with 5 petals
⊞	Radially symmetrical flowers with more than 5 petals
⊞	Bilaterally symmetrical flowers
⊡	Ligneous (woody) plant
▽	Plant which should be protected
▼	Plant protected by law in Britain but not necessarily protected in Europe
▬	Weedy and uncultivated areas
▬	Grasslands
▬	Wooded areas
⌇	Wetlands
◣	Pioneer communities
☠	Poisonous
☠	Slightly poisonous, or might be poisonous

Habitats

Weedy and uncultivated areas

Fields, gardens, waysides, embankments, wasteland, fallow fields, rubbish tips.

Cultivated land offers very specific living conditions. The original vegetation has been removed, the soil has been broken down to make it loose, and the ground is enriched, fertilized and sometimes watered. On such soil wild plants quickly establish themselves. These plants are commonly called weeds, and are characterized by their vigorous growth and rapid multiplication. They reproduce not only by the production of many seeds, but also by means of shoots and runners both above and below ground. Many of these plants have adapted to the abundant fertilizer in the ground and have become nitrogen indicators: they thrive only where the soil contains considerable quantities of nitrogenous salts, as well as an abundance of phosphate and potash. Furthermore, in order to thrive they need relatively open ground with little shade and loose soil.

The kind of soil to some extent determines the selection of weeds which will grow in any particular field, as does the abundance of lime or silicic acid in the subsoil. Years of cultivation with planned applications of fertilizer and herbicides can, however, alter the variety of species to be found anywhere.

Cultivation, beyond steps taken to nurture individual plants, creates conditions appropriate to different groups of weeds. For example, fields of root-crops and potatoes harbour different types to those found in grain fields.

The wild plants found in gardens, which are hoed frequently, are similar to those found in fields of root-crops. However, the variety of cultivated plants found in garden beds has created the basis for some very special weeds to establish themselves.

The dry stone walls which enclose some of these highly cultivated areas are a special habitat, and rock plants take root in the nooks and crannies.

In addition to the plants that grow well on loose soil, there is a second group, which also like nitrogen and fall within the category of synanthropic plants, but which grow on compacted ground. Beaten tracks, paths, waysides, embankments, unused corners, rubbish tips, wasteland, grazing land, even the upper reaches of shores or river banks which are flooded (and therefore fertilized) at most once a year are all typical habitats for ruderal flora (plants that grow on waste ground). Where cultivated land has become neglected there is often a temporary wild growth of annual plants until they give way to the perennials and shrubs which grow very tall, thus shading the ground.

The nutrient content of the soil is the precondition for the occurrence of these plants. Where a large number ensure their survival by means of poisons or thorns, there are usually some hardy strays from far-flung continents such as America and Asia. These have developed from previously introduced garden plants which have since become wild. Species spread along rivers, railway lines and major roads. What is currently of particular interest is the progress of various halophytes (plants which thrive on soil rich in sodium chloride) from the seashore along the motorways and major roads that lead inland and which are salted in winter.

OPPOSITE
Top: *An uncultivated field with an abundance of weeds, a rare sight these days.*

Bottom: *Thistles grow on piles of rubble in abundance, and in a range of species including some rarities.*

Grasslands

**Meadows, grazing land, pastures,
hillsides, turf.**

The prime examples of grassland are the steppes of Asia. This type of land falls between forest and desert. Grasses, with the occasional plant with magnificent blooms, form the main element in the relatively dense vegetation. Plants on the steppes are almost invariably equipped to cope with drought and their annual development is marked by the change from the rainy seasons to longer periods of drought.

In Europe rainfall is far more frequent and not limited to any one season. Meadows and pastures which are used commercially form the greater part of our grassland and the only thing they have in common with the steppe is the high proportion of grasses. In places they are located where previously woods grew and only cutting or animal-grazing ensure that the trees do not begin to grow again.

The ideal meadow for grass for hay-making is the fresh, rich meadow. 'Fresh' refers to the wetness of the ground. Ground can be 'dry', 'fresh', 'damp' or 'wet'. Freshness guarantees that even in years with low rainfall there will be a good grass yield, and in wet years the ground can still be worked with heavy machinery. 'Rich' describes the level of nutrients in the ground. The opposite type of ground is called 'meagre'.

Depending on the degree of dampness and nutrient levels of the soil, and the height above sea level of the location, the flowering plants found in meadows will differ, and with them the nature of the rich meadow. The same meadow can look very different throughout the course of one year: the plants growing there will bloom at different times and so the predominant colour will vary.

The preferences of animals that graze the meadows determines the dominance of certain plants. Fodder which is poisonous, tough or thorny is left and, therefore, increases at the cost of edible plants. If meadows which were previously mown become pasture, many plants that cannot survive being trampled on will disappear. A typical feature of pastureland is the high proportion of trample-resistant white clover.

There is a smooth transition from the types of plants found on wet pasture grounds and damp meadows to the species found in marshes and aquatic areas.

Where the ground is drier with a low level of nutrients, the tall meadow grass gives way to green turf which is usually mown only once a year and is described as 'meagre', 'dry' or 'semi-dry'. The only thing it has in common with garden lawns is the low growth, which in gardens is achieved by frequent mowing. The natural counterpart of the garden lawn is the pasture, which is meagre, sparse meadow for grazing small animals.

The turf found in hilly areas is interspersed with stones and rocks. Here the grass is still the main feature. It is also fertilized by cattle. The meagre grassy areas deteriorate into areas where boulders dominate. Where there are streams and rivulets the ground changes and becomes more marshy. In the higher hills and mountains, in particular, the range of habitats becomes a closely interwoven patchwork.

Grassy areas of limited size or which occur as verges usually have a certain proportion of weeds. These areas are called balks. They resemble grassland when they are mown. Where they are burnt down they tend to form wasteland. Left alone they soon form hedges and woody areas.

Wooded areas

Deciduous, mixed and evergreen woods, lowland woods, woods on slopes, heather, bushes, hedges, wood fringes and clearings.

Without human intervention much of our land would still be woodland. Our forebears cleared land to make fields, meadows and settlements. Only a fraction of the original area is still covered by woods, and these are often intersected by major roads. Moreover, natural woodland is now tended and its spread is predetermined so that it is 'cultivated' woodland, a natural unit planned and managed according to economic criteria.

Plants and bushes which grow across the woodland floor are mostly unaffected by artificial fertilizer and pesticides. The main factors influencing their development are climate, soil and light conditions. In this respect, deciduous woods offer far better growing conditions than evergreen woods, which are permanently steeped in deep shadow. Among deciduous trees which start producing their leaves in May, the ground plants have a good opportunity of growing in full sunlight.

Environmental conditions affect which tree species will be present in an area of woodland. Soil type and drainage, in particular, are often important factors in determining the overall type of wood. On chalk Beech is often the dominant species while on limestone there may be Ash woods. Waterlogged fens can support Alder woods and in riverside woods Willow, Poplars and Alders are common. Richer, dryer soils support a large number of species forming mixed woods, though Oak is usually the dominant tree.

Climate and altitude also influence the type of woodland. At higher altitudes or in colder areas Birch, Hazel and Rowan become more frequent and important. Coniferous trees flourish in these areas too but are also common in warm, dry areas such as southern heathlands where pines, which can withstand the dry, are often the dominant species.

The trees themselves exert a considerable effect on the local environment within the wood and each woodland type has its own characteristic flora. Beech leaves are very slow to rot and Beech woods develop a deep litter of dry fallen leaves which discourages growth of the flowers on the woodland floor. Oak leaves rot faster so oak woods generally have a more prolific ground flora. The practice of coppicing Oak–Hazel woods also encourages the growth of woodland flowers, as regular coppicing lets in extra light to reach ground level.

The light conditions prevailing in bush areas are much better than in the woods, but the remaining habitat factors are worse. Heaths form where the climate is too harsh for tree growth, or the soil is too sandy and too peaty. When the ground is covered by only small, low-lying shrubs this is referred to as dwarf bush heath. Where taller individual bushes rise above the dwarf growth, conditions are somewhat improved.

Hedges are an equivalent of bush areas which have conditions encouraging tree growth. The transitional state is preserved rarely by natural occurrences or more often by occasional clearing, cutting or burning. Compared with wild hedgerows, artificially trimmed hedges are uninteresting as a habitat for ground plants. The woodland margin is a particular type of hedge, nurtured by the forester to provide a wind break and prevented from spreading into nearby fields by the farmer.

Clearings also count as a woodland habitat. The first plants to establish themselves there are usually the annuals which are forced out of the wood by shrubs. The plants in the clearing have adapted so well to these conditions that few of them are to be found outside such areas. Occasionally, although rarely, weeds that have come in from the fields can be found in clearings.

OPPOSITE
Top: *On sparse heaths of dwarf bushes the prevailing species is often heather.*

Bottom: *In woods where light can penetrate ground flora predominate.*

Wetlands

**Water, shores, riverbanks, marshland,
wet meadows, moorland.**

The basic form of non-flowing water is the lake; ponds are usually the result of man-made dams; pools are small and tend to dry out periodically. Often it is hard to distinguish between these three.

Lakes can be divided into three types. Because of the presence of humus materials, moorland lakes are brown in colour and the water is poor in nutrients. There are two types of clear-water lakes: that which is poor in nutrients (often good drinking water) and that which is rich in nutrients, where there is an abundance of plant life. There is always the danger that the latter type will 'turn': too much plant growth will mean that in autumn, when the plants die off, there are large quantities of rotting matter in the water. The result is a lack of oxygen and this can poison the life in the lake.

It is possible to isolate growth zones in lakes. Far out in deep water there are the underwater plants. Then there is a zone of aquatic plants. Further towards the shore are the reeds which give way to a broad band of sedge. The aquatic plants within these zones can sometimes reproduce to such an extent that they temporarily cover the surface of the water.

Running water – streams and rivers – usually displays similar flora to the lake, but not in such great quantities and mainly over a smaller area. Only in quiet coves, where the water flows sluggishly, are conditions similar to those on the lake.

Individual trees or bushes may be found along the shores of the water. Where the shore rises away from the water there is a smooth change in the vegetation from those plants which like the moisture to the drier-ground species. If the shores are flat, a wide marshy area can establish itself. The water is then so close to the surface that it will seep out when the ground is stepped on and remain lying in puddles in the footsteps. With the right kind of subsoil, marshland can occur some distance from lakes or ponds. When the ground water is stagnant or flows very slowly, the species of plants found are different to those growing in fresh water. Marshland does not have to be made up of grasses, sedge or rushes. Woods can also grow there and are referred to as brush, or river-meadow woods.

Flat moors are bogs which have arisen out of lakes. Every lake silts up eventually, even if the process takes many thousands of years. Flat moors are characterized by a high ground water level and high nutrient content.

High moor can originate as flat bogland but usually it forms independently in areas with heavy rainfall. These moors are usually made up of sphagnum, which occurs where the upper part of the plant continues to grow while the lower part dies off and becomes peat. An insulating layer of peat forms between the subsoil and the surface of the high moor. The plants can extract neither nutrients nor water from the ground. They rely on rainwater and dust. High moor habitats are very poor in nutrients.

High moor and flat moor differ greatly from one another, and so does their vegetation. It would be better therefore not to use the same term for both. The word moor describes only the high moor, and marsh or fen is used here for the flat moor.

The transition from marshland to meadow or grassland is a smooth one, and between the two stages there is the wet meadow. Wet habitats can be found anywhere, not only in grassland and woody areas. The only requirement is an abundance of water.

Even though plants cannot survive without water, there are only a few that can cope with a continuous excessive supply. It has a negative effect on the ventilation of the ground and thus on the roots, thereby restricting the plants' breathing. Aquatic plants usually get oxygen to their roots by means of an internal air-pipe system.

OPPOSITE
Top: *Iris are characteristic flora of marshland areas.*

Bottom: *Isolated lakes provide unspoiled habitats for water-lilies and reeds.*

Pioneer communities

Rocks, boulder heaps, scree, steep stony slopes, pebbles, sand, dunes, slag, walls.

Whenever soil, particularly humus, is found only in extremely small amounts, this is the habitat of pioneer plants. Their demands are small and yet they manage to produce large growth. When they die off, they sometimes form the basis for some more demanding plant. These first settlers in habitats alien to vegetation rarely grow in a mass, but are usually found growing sparsely with great gaps between them. This is very evident on rocks where minimal quantities of water and minerals have gathered in the cracks, thereby providing a basis for plant life. With an excretion from their roots, the plants manage frequently to loosen the stony surface still further and thus extend the area in which they can survive. This does, however, lead to weathering and destruction of the rocks. At the start a plant's chances of taking root are very slim. Larger patches of plant life find an opportunity to grow along the wide cracks in the bands of rock or on horizontal ledges, where life-supporting materials can gather.

In the rocks and boulder masses the cushion and trellis plants predominate. These are half-spherical with all their elements huddled together for protection, or spread out but clinging closely to the ground. The picture changes on scree covering ground below rocks, where the roughened chunks can be found on top of each other, each one, on average, the size of a fist. Scree is unstable, and stones are constantly slipping away. Blocks of rock above the scree crumble and tumble down. The plants that survive best are those that have plenty of runners and shoots which simply grow through every new fall of rock. They become a network that eventually enables the mobile mass of rock to establish itself and become stable.

OPPOSITE
The flower world in rocky Alpine areas is varied and colourful.

Coarse soils where fine crumbs of earth are few and humus content is low can be divided into two categories: moving and sedentary soils. The size of the crumbs is of little relevance. Whether we are dealing with coarse or fine screes, pebble or slag heaps, or dunes, moving and sedentary soils can be recognized by their vegetation and its growth. The general rule is that sedentary soils produce denser growth and show more resemblance to wastelands or grassland vegetation, except that the plants that grow there demand less humus and nutrients.

Further changes in the vegetation in these habitats are caused by the mineral composition of the coarse material. Limestone and silicates offer different conditions as rock, scree and sand. In the case of limestone, useful nutrients effloresce from the rock itself. It can happen that two types of one species are only marginally different in their form, but one grows on limestone and the other on acid rock.

On steep slopes the ground is rapidly washed away so that the only soil available is undeveloped and 'mountain fresh'. It does not contain any humus and is not a good tilth, so the kinds of plants that tend to grow on it are pioneer plants. These experience the same difficulties in becoming established as they would on scree.

Identifying plants

Flower colour

Decide between white, yellow, red, blue, green or brown. Where flowers have several colours, choose the dominant one. A white flower with red dots will be found in the white flower section. In 90 per cent of cases establishing flower colour is easy, but there are two cases where opinions differ as to which colour dominates. Since the colour of the flower may vary slightly in shade from specimen to specimen it is wise to look up both colour sections. The borderline cases are (1) red-violet/blue-violet; and (2) light yellow/creamy white/pale green.

Flowers that have just started blooming often still bear the greenish bud colour, or have not quite attained their full colour. Flowers that are dying sometimes fade in colour. It is therefore wise to go by the colour of the flower in bloom.

In cold weather the petals often turn red or blue in the bud stage. Sometimes this is just a hint of colour, at other times it is a very intense colouring. The colour change is, however, always restricted to the outside, or underside, so always go by the inner surface.

Flower form

Next decide on the form. The flower will be either clearly bilaterally symmetrical (flower can be divided into identical halves along one line of symmetry only) or more or less radially symmetrical (flower is circular and can be divided into identical halves along any diameter). In the latter case, count how many petals the flower has, or, if it has none, how many petal tips the corolla displays (the corolla is the collective name for the petals).

The radially symmetrical flowers are divided into three groups: those with up to 4 petals, those with 5 petals, and those with more than 5 petals.

Radially symmetrical flowers with up to 4 petals. There are flowers with 3 or 4 petals (2 petals is very rare). Sometimes the petal tips themselves are somewhat unevenly distributed (Speedwell, Mint, for example) so that strictly speaking the flower is not entirely radially symmetrical. Plants with this kind of flower are listed in this section because the amateur will hardly recognize the bilateral symmetry on small flowers.

Radially symmetrical flowers with 5 petals. Here it should be remembered that certain flowers have 5 split or deeply divided petals (Splendid Pink and Lesser Stitchwort respectively) which might on superficial examination resemble flowers with many more petals.

Radially symmetrical flowers with more than 5 petals. This group includes many flowers with more than 6 petals, some even with 7 or 8. The family Asteraceae (Compositae), such as the daisy is included as the layman generally considers the capitulum (a group of flowers on a single stem in the form of a disc), which is made up of tiny florets, as one flower head. Those with more experience should therefore look up Asteraceae (Compositae) in this particular group.

There is only one group of **bilaterally symmetrical flowers.** Here the number of petals is irrelevant. The corolla (part of the flower formed by the petals) is usually bell-shaped or features an upper or lower lip and is thus clearly identifiable as bilateral. These include campanulate (bell-shaped), papilionaceous (butterfly-shaped), labiate (lipped) and violaceous (pansy-shaped) flowers.

Woody plants. Only a selection of ligneous (woody) plants is listed. They are mostly grouped together at the end of each colour block.

Pictorial key and plant identification

The pages on which each classification can be found are given in the contents list on p.3, or, even more easily, by using the key that follows on the next 8 pages, where you will find photographs of 6 flowers which are typical of each of the flower form groups:

p.25: radially symmetrical flowers with up to 4 petals

p.27: radially symmetrical flowers with 5 petals

p.29: radially symmetrical flowers with more than 5 petals

p.31: bilaterally symmetrical flowers.

The important point being illustrated in the photographs is flower form, rather than flower colour. Opposite each set of photographs is a 6-part colour band.

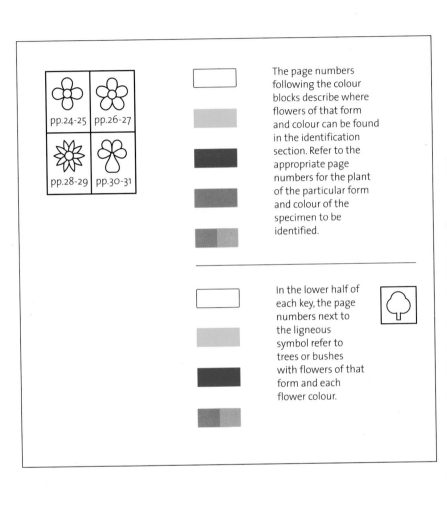

The page numbers following the colour blocks describe where flowers of that form and colour can be found in the identification section. Refer to the appropriate page numbers for the plant of the particular form and colour of the specimen to be identified.

In the lower half of each key, the page numbers next to the ligneous symbol refer to trees or bushes with flowers of that form and each flower colour.

Radially symmetrical flowers with up to 4 petals

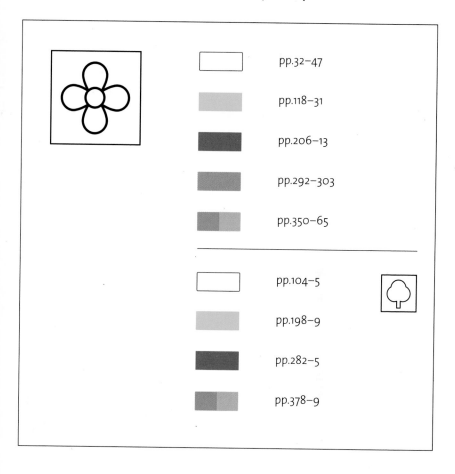

pp.32–47

pp.118–31

pp.206–13

pp.292–303

pp.350–65

pp.104–5

pp.198–9

pp.282–5

pp.378–9

Examples of radially symmetrical flowers with up to 4 petals

Radially symmetrical flowers with 5 petals

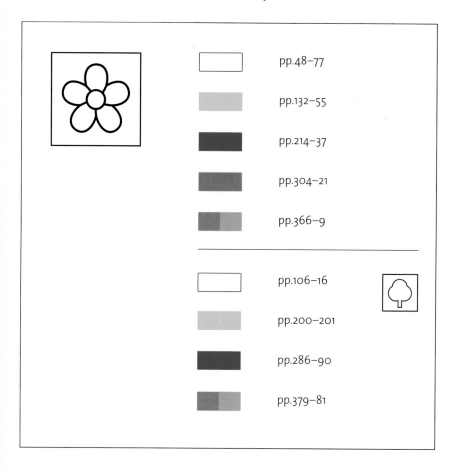

pp.48–77

pp.132–55

pp.214–37

pp.304–21

pp.366–9

pp.106–16

pp.200–201

pp.286–90

pp.379–81

Examples of radially symmetrical flowers with 5 petals

Radially symmetrical flowers with more than 5 petals

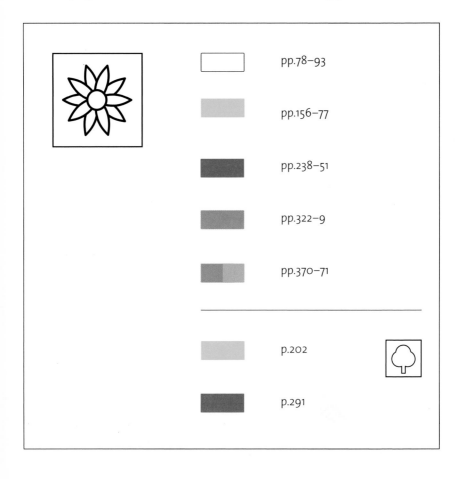

pp.78–93

pp.156–77

pp.238–51

pp.322–9

pp.370–71

p.202

p.291

Examples of radially symmetrical flowers with more than 5 petals

Bilaterally symmetrical flowers

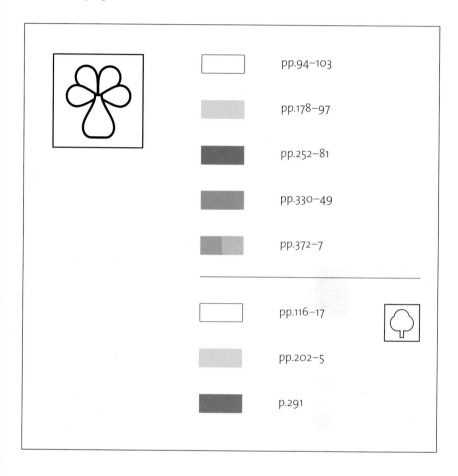

pp.94–103

pp.178–97

pp.252–81

pp.330–49

pp.372–7

pp.116–17

pp.202–5

p.291

Examples of bilaterally symmetrical flowers

Water Plaintain
Alisma plantago-aquatica

Arrowhead
Sagittaria sagittifolia

Water Plaintain family *Alismataceae*
Flowering time: June–Aug.
Height of growth: 20–100 cm
Monocotyledonous; Perennial

Identification marks
Whorled panicles. Trimerous flowers, petals withering very rapidly, white or pink, yellowish base. Leaves oval, robust, on long stalks, forming basal rosette.

Habitat and distribution
Margins of stagnant or slow-flowing waters, in reed beds and sedge, also in ditches. Frequently indicator of muddy ground rich in nutrients. Throughout most of Britain.

Additional information
Similar to some closely related but much rarer species, which are distinguished by their broad leaves and their habitat: Narrow-leaved Water Plantain (*Alisma lanceolatum*), narrow leaves; Ribbon-leaved Water Plantain (*Alisma gramineum*), leaves ribbon-like, submerged, protected species; both aquatic plants.

Water Plantain family *Alismataceae*
Flowering time: July–Aug.
Height of growth: 30–90 cm
Monocotyledonous; Perennial

Identification marks
Erect stem, triangular in cross-section with whorled panicles. Flowers large, up to 2 cm wide, unisexual, the upper ones male, the lower ones female (usually with noticeably shorter stalks). Base of petals has violet tinge. Leaves all basal, some ribbon-like and submerged, some long-stalked, erect with arrow-shaped blade.

Habitat and distribution
In reed-beds in still or slow-flowing waters. Likes water rich in nutrients. Local in England, rarer in the North.

Additional information
The shallow water version develops ribbon-like leaves and petiolate leaves; in deep water only the ribbon-like leaves are present. These are absent on the (stunted) land version.

Water Soldier
Stratiotes aloides

Common Frogbit
Hydrocharis morsus-ranae

Frogbit family *Hydrocharitaceae*
Flowering time: June–Aug.
Height of growth: 15–50 cm
Monocotyledonous; Perennial

Identification marks
Leaves robust, stiff, triangular in cross-section, long and narrow, margin with spiny teeth. They form a thick funnel-shaped rosette which is usually submerged. Many offshoots. Flowers very large, up to 4 cm wide, males on long stems coming from leaf axils, encircled by a white spathe, females sessile.

Habitat and distribution
In lakes, still creeks, ponds and ditches. Mainly floating in the water which must be rich in nutrients but deficient in chalk. Scattered and very local, mainly in eastern England.

Additional information
Male and female flowers on separate plants. Only female flowers produced in Britain.

Frogbit family *Hydrocharitaceae*
Flowering time: July–Aug.
Height of growth: 15–30 cm
Monocotyledonous; Perennial

Identification marks
Plant which floats on water surface. Leaves rounded and long-stalked, heart-shaped at base of stalk, tough, leathery. Flowers one-sexed; males usually in threes, long-stalked, approx. 3 cm wide; female smaller, solitary, short-stalked.

Habitat and distribution
In still or slow-flowing waters and in ditches. Likes shady water, low in chalk and not too cool. Scattered throughout England and Wales, sometimes locally common; absent from Scotland.

Additional information
When not blossoming, similar to even rarer Fringed Water-lily (*Nymphoides peltata*, flowers yellow, in clusters of 5). However on Frogbit at the base of each leaf-stalk there are 2 large, brownish, oval stipules.

Lords and Ladies
Arum maculatum

Bog Arum
Calla palustris

Arum family *Araceae*
Flowering time: April–May
Height of growth: 30–45 cm
Monocotyledonous; Perennial

Identification marks
Greenish-white spathe encloses the spadix. Male flowers are above the female. Spadix becomes club-shaped at the top, emits smell of carrion. This feature serves as a fly-trap for the purpose of pollination. Leaves sagittate long-stalked, arrow-shaped and sometimes somewhat spotted.

Habitat and distribution
England and Wales, rarer in Scotland. On loose soil rich in nutrients in deciduous or mixed woodland, in thickets and along hedgerows. Prefers not too dry, loamy soil rich in mull. Likes warm situations.

Additional information
Colour of club can vary from white to violet. Other common names: Cuckoo-pint, Jack-in-the-pulpit.

Arum family *Araceae*
Flowering time: June–July
Height of growth: 15–30 cm
Monocotyledonous; Perennial

Identification marks
Creeping rhizome below and above ground. Leaves stalked, heart-shaped, leathery, shiny. Flowers in spadix terminating in club-like apex at base of which is large striking spathe. Spadix has coral red berries.

Habitat and distribution
In swamps and wet woodlands. Introduced over 100 years ago and now naturalized in a few places in S. England.

May Lily
Maianthemum bifolium

Baneberry
Actaea spicata

Lily family *Liliaceae*
Flowering time: May–June
Height of growth: 8–20 cm
Monocotyledonous; Perennial

Buttercup family *Ranunculaceae*
Flowering time: May–June
Height of growth: 30–65 cm
Dicotyledonous; Perennial

Identification marks
Usually only 2, stalked, heart-shaped leaves on stem. These are leathery, with entire margin and parallel venation. Flowers small, in a terminal raceme. Fruit globular, red shiny berries (barely 0.5 cm in diameter).

Habitat and distribution
Shady woodland, likes thick moss growing over at least surface acidity, low nutrient loamy soil with thick thoroughly decomposed humus layer. Very rare. Only in a few localities in England.

Identification marks
Straight stem, few branches. Large bi-pinnate leaves with 3 leaflets; when rubbed they give off unpleasant smell. Leaf is divided into hand-shaped toothed segments, margin spiny serrate. Flowers small, in terminal and lateral racemes. Leaves usually extend beyond them.

Habitat and distribution
Dense thickets and shady deciduous woodland (especially of Ash), and limestone pavements. Likes porous, well-drained limestone soils. Local in parts of northern England.

White Alpine Poppy
Papaver sendtneri

Garlic Mustard
Alliaria petiolata

Poppy family *Papaveraceae*
Flowering time: July–Aug.
Height of growth: 5–20 cm
Dicotyledonous; Perennial

Identification marks
All leaves in basal rosette, asymmetrically pinnate with usually 2 pairs of oval lobes. Flowers solitary on straight stems with no branches; pedicels hairy; flowers approx. 4 cm wide, 2 sepals wither rapidly. Fruit is hairy oval capsule with star-like apex, approx. 1 cm long.

Habitat and distribution
Not British. Only in Alpine areas. On scree and rubble in calcareous Alps, rare. Seldom found under 1,800 m.

Additional information
This plant belongs to a whole group of closely related and similar species which grow both on chalk and on gravel scree in Alpine areas. Flowers white, yellow or red.

Mustard family *Brassicaceae (Cruciferae)*
Flowering time: April–June
Height of growth: 20–120 cm
Dicotyledonous; Biennial

Identification marks
Erect stem, leaves alternate, flowers in a false umbel. Lower leaves long-stalked, kidney- or heart-shaped, coarsely crenate. Upper leaves heart-shaped to ovoid, broadly toothed. When rubbed all parts of the plant smell strongly of garlic. Fruit is long pod.

Habitat and distribution
Hedgerows, margins of woodland, wayside verges, scree, rubble, walls, fences. Likes nitrogen, needs humidity. Prefers loose, well-drained soils rich in humus. Frequent throughout most of Britain, rare in higher locations.

Additional information
Also called Hedge Garlic and Jack-by-the-hedge; scientific name was *A. officinalis*.

Sea Rocket
Cakile maritima

Seakale
Crambe maritima

Mustard family *Brassicaceae (Cruciferae)*
Flowering time: June–Aug.
Height of growth: 15–30 cm
Dicotyledonous; Annual

Identification marks
Plant glaucous. Stalk branched, ascending. Leaves fleshy, undivided or deeply lobed, alternate. Flowers scented, a good 0.5 cm wide with approx. 0.5 cm long, nectar-filled tube. Fruit characteristically of two sections, upper mitre-shaped, lower oval-shaped. Section between has pointed protruberances.
Habitat and distribution
Occurs along coast, on sandy beaches and in dunes. Can bear high salt concentrations but does not need them to thrive. Rarely spreads inland and does not survive well there.
Additional information
Plants occur with lilac, purple or white flowers.

Mustard family *Brassicaceae (Cruciferae)*
Flowering time: June–Aug.
Height of growth: 40–60 cm
Dicotyledonous; Perennial

Identification marks
Very branched stem. Lower deciduous leaves large, lobed and wavy, long-stalked, becoming smaller towards the top, the upper ones narrow, linear, short-stalked. Numerous flowers in dense branched clusters. Flowers 1–1.5 cm wide. Fruit is small globular pod. Entire plant glaucous.
Habitat and distribution
On sand or between rocks on beach and on primary dunes as far north as Central Scotland. Likes salt; survives flooding and burial in shingle.
Additional information
It is occasionally grown as a garden vegetable.

Wild Radish
Raphanus raphanistrum

Hare's-ear Mustard
Conringia orientalis

Mustard family *Brassicaceae (Cruciferae)*
Flowering time: May–Sept.
Height of growth: 20–60 cm
Dicotyledonous; Annual

Identification marks
Erect stem, relatively branched, stiff hairs. Leaves pinnately lobed, upper ones entire, lanceolate. Flowers white or yellow with yellowish or deep violet veins. Sepals erect. Fruit narrow cylindrical; constricted strongly between seeds.

Habitat and distribution
Common in fields, also on rubble and in gardens. On light and heavy soil, likes it somewhat acidic.

Additional information
Plant bears similarities, especially in its yellow version, to Charlock (*Sinapis arvensis*, p. 120), which often grows in the same areas but has horizontal sepals.

Mustard family *Brassicaceae (Cruciferae)*
Flowering time: May–July
Height of growth: 10–50 cm
Dicotyledonous; Annual

Identification marks
Hairless plant, glaucous. Stalk erect, usually unbranched. Leaves elliptic-ovoid, flattened, clasping stem. Flowers large, in loose umbel; fruit stalked, and spreading, curved and 4-angled.

Habitat and distribution
In fields, along wayside and in uncultivated fields. Likes soil not too wet, rich in nutrients and chalk. Casual, quite frequent but rarely becoming established.

Additional information
Similar: Tower Mustard (*Arabis glabra* = *Turritis glabra*), another rare introduction. Leaves are pointed and pods are all twisted to one side and down-curved.

Hoary Cress
Cardaria draba

Mustard family *Brassicaceae (Cruciferae)*
Flowering time: May–June
Height of growth: 30–90 cm
Dicotyledonous; Perennial

Identification marks
Stem usually branched in inflorescence. Lower leaves stalked, remainder clasping stem, margin entire or serrated. Flowers scented; dense inflorescence. Fruit long-stalked, heart-shaped to globular-inflated pod.

Habitat and distribution
Arable land. Prefers exposed chalk areas rich in nutrients in dry warm locations. Throughout England and Wales, spreading.

Additional information
Often included in Cress family (*Lepidium*) and has many similarities. Its pods, however, are usually heart-shaped and not split at the tip.

Common Whitlow-grass
Erophila verna

Mustard family *Brassicaceae (Cruciferae)*
Flowering time: March–June
Height of growth: 3–9 cm
Dicotyledonous; Annual

Identification marks
All leaves in basal rosette, lanceolate, margin entire or with 4 teeth. Inflorescence racemose. Petals indented deeply. Fruit is a flat, elliptical pod.

Habitat and distribution
On rocks, walls, gravel and railway slag, also among sparse dry turf and semi-dry turf. Mainly spread over warm, not too damp sandy soil with sufficient nutrients. Common, but easily overlooked.

Additional information
Similar: Shepherd's Cress (*Teesdalia nudicaulis*); strong leaves in rosette shape, pinnatifid. Flower petals of unequal size; throughout Britain. Thale Cress (*Arabidopsis thaliana*), fruit long and thin; in similar habitats throughout Britain.

Cuckoo Flower
Cardamine pratensis

Hairy Bitter-cress
Cardamine hirsuta

Mustard family *Brassicaceae (Cruciferae)*
Flowering time: April–June
Height of growth: 15–60 cm
Dicotyledonous; Perennial

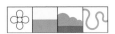

Identification marks
Basal leaves form rosette, pinnate. Leaflets rounded, terminal leaflet usually larger. Stem leaves pinnate with narrow tips. Hollow stem. Inflorescence a raceme. Flowers large (approx. 1–1.5 cm across), petals longer than sepals. Fruit much longer than wide. Flower colour depends on habitat: white (shady), pink, mauve or deep purple (dry).

Habitat and distribution
Mainly in damp meadows and pastures on loamy soil. Indicator of rich ground and ground water. Common.

Additional information
Similar: Narrow-leaved Bitter-cress *(C. impatiens)*, flowers only 0.5 cm across, petals as long as sepals, whitish; fruit bursts if touched; woods.

Mustard family *Brassicaceae (Cruciferae)*
Flowering time: April–Aug.
Height of growth: 7–30 cm
Dicotyledonous; Annual

Identification marks
Basal rosette of pinnate leaflets. Stem usually greatly branched often from base, erect, few or no leaves. Flowers small, under 0.5 cm across, petals considerably longer than sepals. Fruit long thin pod.

Habitat and distribution
Bare ground, walls, gardens, newly laid lawns, waysides, hedgerows. Prefers sandy soils, low in chalk but rich in nutrients and not too dry. Likes half-shade. Common.

Additional information
Similar: Wavy Bitter-cress (*C. flexuosa*), more robust, at least 5 leaves per stem. In damp shady places throughout Britain.

Large Bitter-cress
Cardamine amara

Watercress
Nasturtium officinale

Mustard family *Brassicaceae (Cruciferae)*
Flowering time: April–June
Height of growth: 10–60 cm
Dicotyledonous; Perennial

Identification marks
Stalk filled with pulp, slightly pentagonal, usually ascending. Leaves simple or double pinnate. Racemose inflorescence. Flowers large, approx. 1 cm diameter. Anthers purple violet.

Habitat and distribution
In streams and ditches, fens and other wet places. Likes wet soil rich in nutrients. Somewhat calcifugous. Scattered throughout Britain up to heights of about 500 m in Scotland. Woodland in N. Europe.

Additional information
Similar, and often confused with: True Watercress (*Nasturtium officinale*, see right), with hollow stem and yellow anthers. Makes sharper but not so bitter 'wild' salad.

Mustard family *Brassicaceae (Cruciferae)*
Flowering time: May–Oct.
Height of growth: 10–38 cm
Dicotyledonous; Perennial

Identification marks
Stem hollow, ascending, lower part often prostrate with roots; sometimes submerged. Very lowest leaves divided into 3 leaflets, upper leaves alternate, pinnate, terminal leaflet usually larger. Flowers in clusters; anthers yellow.

Habitat and distribution
In springs, ditches and streams with clear, cool, fast-flowing water. On sandy or muddy ground rich in nutrients. Common throughout lowland Britain but becoming less common as a result of water pollution.

Additional information
One-rowed Watercress (*N. microphyllum*), more delicate, has longer pods; greenery tinged red-brown in winter; Watercress (*N. officinale*) remains deep green all year round.

Shepherd's Purse
Capsella bursa-pastoris

Field Penny-cress
Thlaspi arvense

Mustard family *Brassicaceae (Cruciferae)*
Flowering time: All year
Height of growth: 7–45 cm
Dicotyledonous; Annual – Biennial

Identification marks
Rosette of long dentate basal leaves. Stem branched. Upper leaves margin entire, clasping stem. Flowers in loose racemes. Fruit triangular, slightly heart-shaped, flat, projecting erect. Stalks spreading.

Habitat and distribution
Grows well on soils not too dry, not too shady but rich in nutrients. In gardens, fields, on grassy areas, embankments, along paths, on waste ground. Common throughout Britain.

Additional information
Through self-pollination many local populations have occurred, so variations common from area to area.

Mustard family *Brassicaceae (Cruciferae)*
Flowering time: May–July
Height of growth: 10–60 cm
Dicotyledonous; Annual – Biennial

Identification marks
Stem erect, angular, often branched at top. Stem lanceolate, leaves usually broadly toothed, with sagittate base. Flowers small, initially racemose, elongating greatly in fruit. Fruit over 1 cm wide, flat, round, broadly winged, with a deep notch apically. When rubbed the plant smells strongly of leek.

Habitat and distribution
Arable land, gardens, waysides and rubbish dumps. Frequent. Likes loose loamy soil rich in nutrients.

Additional information
Similar: Perfoliate Penny-cress (*T. perfoliatum*), more delicate, stem smooth, leaves usually have margin entire. Likes calcareous soils and confined to a few places in Midlands, though occurs as a casual elsewhere.

Round-leaved Penny-cress
Thlaspi rotundifolium

Water Chestnut
Trapa natans

Mustard family *Brassicaceae (Cruciferae)*
Flowering time: July–Sept.
Height of growth: 5–15 cm
Dicotyledonous; Perennial

Identification marks
Stem creeping, producing erect or ascending flowering shoots which bear many leaves. Leaves glaucous, ovoid, margin entire or serrate. Flowers in compressed corymbs. Fruit oval, somewhat flattened, slightly winged at edge, approx. twice as long as wide.
Habitat and distribution
Not British. Only in Alpine areas over 1,000–1,500 m.
Additional information
Prevalent form is distinguished by its blue-tinged corolla with darker veins. White-flowered form is rare.

Water Chestnut family *Trapaceae*
Flowering time: July–Aug.
Height of growth: 50–300 cm
Dicotyledonous; Annual

Identification marks
Stem submerged, floating rosette of many rhomboid, leathery and spreading leaves. Longer leaf-stalks have hollow swelling below the blade. Single flowers in leaf axil, easily overlooked, very perishable. Fruit plus woody sepals forms thorn-bearing 'nut' 2–4 cm wide.
Habitat and distribution
Not British. Mostly planted. In stagnant waters rich in nutrients, low in chalk. Rare, but where it occurs it is usually in large numbers. Often planted but rarely thrives.
Additional information
The many forms of this now widespread plant are distinguished by the shape of the nut.

Enchanter's-nightshade
Circaea lutetiana

Thyme-leaved Speedwell
Veronica serpyllifolia

Willow-herb family
Onagraceae (Oenotheraceae)
Flowering time: June–Aug.
Height of growth: 20–70 cm
Dicotyledonous; Perennial

Figwort family *Scrophulariaceae*
Flowering time: March–Oct.
Height of growth: 10–30 cm
Dicotyledonous; Perennial

Identification marks
Stem usually upright, often branched, hairy. Leaves opposite, ovoid-lanceolate, dull, hairy. Terminal racemes. Flowers small, sometimes faintly pink. Peduncles without sub-tending leaves.

Habitat and distribution
In all types of woodland, even in clearings, provided the soil is damp enough. Likes heavy loamy soil rich in nutrients. Common; usually in large numbers.

Additional information
Similar: the rarer and more delicate Alpine Enchanter's-nightshade (*C. alpina*), with smooth glossy leaves and bristle-shaped bracteoles under flower stems. More common is the hybrid: Upland Enchanter's-nightshade (*C. intermedia*)

Identification marks
Stalk creeping, ascendent or erect. Leaves round to ovoid, upper ones smaller and narrower, weakly toothed. Flowers single in axils of upper leaves, small, white, with blue-violet veins. One petal smaller than the other three.

Habitat and distribution
Fields, paths, riverbanks, heaths, grassy area communities on meadow paths and pastures. Prefers moist, heavy (compressed), loamy soil, deficient in lime but nitrogenous. Widespread.

Additional information
Subspecies *V. humifusa*, long, creeping, rooting stem; large, blue flowers; occurs locally in damp, mountainous places in Wales, N. England and Scotland.

Buck's-horn Plantain
Plantago coronopus

Sweet Woodruff
Galium odoratum

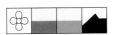

Plaintain family *Plantaginaceae*
Flowering time: July–Aug.
Height of growth: 10–40 cm
Dicotyledonous; Annual

Madder family *Rubiaceae*
Flowering time: May–June
Height of growth: 10–30 cm
Dicotyledonous; Perennial

Identification marks
All leaves in basal rosette, almost entire, coarsely serrate or most usually 1–2 pinnatifid. Usually many flower stems, ascending, with terminal, linear spikes. Flowers very small, stamens extend beyond them.

Habitat and distribution
Seashore, dunes, saline meadows. Sometimes only along the littoral; absent from some inland areas. Likes highly compressed, salty, moist (loamy) soil.

Additional information
Inflorescence is similar on other plantains (see pp.302, 365), but this type has divided leaves.

Identification marks
Stem erect, unbranched, quadrangular in cross-section. Leaves lanceolate to elliptic, dark green, in whorls of 6–8. Terminal cymes with funnel-shaped flowers. Releases characteristic hay smell when plant dying down.

Habitat and distribution
In all woodlands but clear preference for deciduous forests. Common; but only on soils which are porous, not too dry, rich in nutrients and mull. Usually in large numbers.

Additional information
The plant was for a long while classed by systematists as *Asperula odorata*, from which it got its well-known popular name: that of 'Our Lady's Bedstraw' will not replace it.

Wood Bedstraw
Galium sylvaticum

Round-leaved Bedstraw
Galium rotundifolium

Madder family *Rubiaceae*
Flowering time: June–Aug.
Height of growth: 30–100 cm
Dicotyledonous; Perennial

Identification marks
Stem erect, usually branched, round. Leaves linear-lanceolate, bluish-green, in whorls of 6–8. Flowers in loose projecting panicles. Plant often has reddish tinge, especially on stem parts and leaf ribs.
Habitat and distribution
Not British. In deciduous and mixed woodland on calcareous, slightly moist but warm, loamy soil containing mull. Frequent; up to around 1,000 m.

Madder family *Rubiaceae*
Flowering time: June–Sept.
Height of growth: 10–25 cm
Dicotyledonous; Perennial

Identification marks
Stem prostrate to ascending, thin, quadrangular, rarely branched. Leaves always in whorls of 4, oval to round, short hooked bristles. Few flowers, clustered panicles. Fruit made up of two globular parts, hairy.
Habitat and distribution
Not British. Originally in shady coniferous forests in higher altitudes. Likes well-moistened rather poor soils, low in lime and with abundant (acidic) humus layer. Often found in the Alps and Mittelgebirge in Germany, otherwise often introduced with young pine trees (and thrives). Only rare in northern Europe nowadays.
Additional information
In older botanical works the species is still listed under the scientific name of *G. scabrum*.

Cleavers
Galium aparine

Hedge Bedstraw
Galium mollugo

Madder family *Rubiaceae*
Flowering time: June–Aug.
Height of growth: 15–120 cm
Dicotyledonous; Perennial

Identification marks
Stem quadrangular in cross-section, climbs with the aid of hooked prickles. Leaves linear, cuneate, 6–8 in a whorl (help support stem when climbing), rough. Inflorescence in leaf axils, few flowers, longer than the leaves.

Habitat and distribution
In fields, gardens and hedgerows, on rubbish dumps, also in thickets, woodland margins and scrub along river banks. Common. Nitrogen indicator; likes moist loamy soils rich in nutrients.

Additional information
Similar but much less widespread: Corn Cleavers (*G. tricornutum*). Usually its inflorescences comprise only three flowers and are shorter than the leaves. In cornfields; likes soil containing lime.

Madder family *Rubiaceae*
Flowering time: June–Sept.
Height of growth: 25–120 cm
Dicotyledonous; Perennial

Identification marks
Stem ascending or erect, quadrangular in cross-section, smooth or hairy, usually branched. Leaves narrow, pointed, mostly in whorls of 8. Many-flowered panicle. Tips of flowers rounded with fine point.

Habitat and distribution
Meadows, dry grassy slopes, waysides, thickets. Often on rich loamy soil. More or less throughout Britain.

Additional information
Similar: in wetter habitats: Common Marsh Bedstraw (*G. palustre*) with red anthers, Fen Bedstraw (*G. uliginosum*) with yellow anthers. Both have rough, bristled stem. Common. On poor soil low in lime: Heath Bedstraw (*G. saxatile*) on heaths, moors and in grassy areas – the flower tips are pointed but without a prickle.

Pyrenean Bastard Toadflax
Thesium pyrenaicum

Black Bindweed
Fallopia convolvulus

Sandalwood family *Santalaceae*
Flowering time: June–Sept.
Height of growth: 20–60 cm
Dicotyledonous; Perennial

Identification marks
Bushy but without stolons. Stem erect, branched above. Leaves narrow, bluish-green with 3, sometimes 5, veins. Many-flowered panicle; below each blossom 3 small bracts. Small fruit, round-ovoid, short flower remnant at the top.

Habitat and distribution
Not British. Sunny forest margins, open woodland, mountain heaths. Found on rather dry and often stony ground containing lime. Scattered but absent in areas where summer is cool and rainfall heavy.

Additional information
Synonym: *T. montanum.*

Dock family *Polygonaceae*
Flowering time: July–Oct.
Height of growth: 10–100 cm
Dicotyledonous; Annual

Identification marks
Stem thin, angular, bent or climbing up on other plants; stem and underside of leaves are mealy. Leaves stalked, triangular/arrow-shaped. Blossoms triangular, 2–5 in leaf axils and clustered to form a false spike.

Habitat and distribution
Arable fields, wasteland, gardens, thickets. Common. Likes loamy soils, not too dry and rich in nutrients.

Additional information
Similar: Copse Bindweed (*F. dumetorium*), more robust, 1–2 m, stem smooth. Scattered in moist thickets and hedges. Both previously belonged first to the genus *Polygonum* and then *Bilderdykia*.

Pale Persicaria
Polygonum lapathifolium

Knotgrass
Polygonum aviculare

Dock family *Polygonaceae*
Flowering time: June–Oct.
Height of growth: 25–75 cm
Dicotyledonous; Annual

Identification marks
Stem erect or ascending in bends, usually richly branched, with thickened nodes. Leaves oval, broadest over the basal third, often with dark spots. Leaf sheath (ochra) membranous, cornet-shaped, the uppermost with short cilia. Flower spikes at ends of branches, many-flowered, white or rarely pink.

Habitat and distribution
Weedy places in fields, on banks of rivers and ponds, less common along paths or in water. Likes moisture and plenty of nutrients. Common.

Additional information
Similar: Redshank (*P. persicaria*, p.214), all leaf sheaths have fringe of cilia, often grow together in similar habitats. Flowers also white but usually pink.

Dock family *Polygonaceae*
Flowering time: June–Oct.
Height of growth: 3–200 cm
Dicotyledonous; Annual

Identification marks
Stem prostrate, richly branched, ascending from branch bases. Branches bear leaves up to the apex. Membranous leaf sheath embracing stem. Flowers axillary, single or in small numbers, greenish with white or red edge, small, barely 3 mm long, but numerous.

Habitat and distribution
Fields, rubbish tips, wasteland, verges, roadsides. Very common everywhere. Nitrogen indicator.

Additional information
Often overlooked, but cannot be mistaken. An aggregate of different species in differing habitats. The true *P. aviculare* is commonest.

Chickweed
Stellaria media

Lesser Stitchwort
Stellaria graminea

Pink family *Caryophyllaceae*
Flowering time: Jan.–Dec.
Height of growth: 5–40 cm
Dicotyledonous; Annual

Identification marks
Stem prostrate, much branched, round, hairs in single longitudinal row down each internode. Leaves ovoid, opposite, lower ones petiolate. Flowers few in each axil. Petals usually as long as sepals, deeply indented at tip.
Habitat and distribution
Fields, gardens, rubbish tips, path verges, also in woods. Likes well-moistened, nitrogenous (over-fertilized) soils. Very common.
Additional information
A very variable species showing considerable range of size, hairiness, size and number of petals and seed characters.

Pink family *Caryophyllaceae*
Flowering time: May–Aug.
Height of growth: 20–90 cm
Dicotyledonous; Perennial

Identification marks
Stem quadrangular, brittle, ascending or scrambling: climbs with help of out-spread leaves. These are narrow linear-lanceolate, opposite, grass-green. Terminal, branched cymes. The 5 petals cleft almost to the base, about as long as the calyx.
Habitat and distribution
Prefers grassy habitats: meadows, pastures, banks, less common along edges of fields, pathways and thicket margins. Likes calcifugous, acid soils which are not too moist and not too rich in nutrients. Common throughout most of Britain.
Additional information
Similar: *Stellaria longifolia = S. diffusa*. Leaves yellowish-green, rough at the edges; very rare in wet woodlands. Not British.

Greater Stitchwort
Stellaria holostea

Wood Stitchwort
Stellaria nemorum

Pink family *Caryophyllaceae*
Flowering time: April–June
Height of growth: 15–60 cm
Dicotyledonous; Perennial

Identification marks
Stem usually erect growing up from bent base, angular. Leaves opposite, sessile, lanceolate, usually stiff and dark green with rough edges. Branched cymes; flowers large, with leafy bracts. Petals cleft approximately to the middle, 1–1.5 cm long.

Habitat and distribution
Especially in deciduous and mixed woodland, woodland margins and hedgerows. Likes calcifugous acid loamy soil rich in nutrients and mixed sand, but tolerates wide range of mull soils. Common throughout Britain.

Additional information
Distant resemblance; Marsh Stitchwort (*S. palustris*), petals 0.5–1 cm, up to twice as long as calyx, cleft almost to the base. Leaves thickish, glaucous with smooth edges. Marshes; scattered.

Pink family *Caryophyllaceae*
Flowering time: May–June
Height of growth: 15–60 cm
Dicotyledonous; Perennial

Identification marks
Stem weak ascending, round, with soft hairs all round, very brittle. Often has long creeping runners. Leaves opposite, heart-shaped to ovoid, lower ones petiolate. Petals about twice as long as sepals, deeply bifid. 3 styles.

Habitat and distribution
Shady deciduous woods, and streams. Likes marshy ground soaked in ground water and rich in nutrients. Scattered, in the north and west of Britain.

Additional information
Similar: Water Chickweed (*Myosoton aquaticum*), scattered on river banks, in moist woodland, marshes and fens. Flower has 5 styles.

Field Mouse-ear
Cerastium arvense

Common Mouse-ear
Cerastium fontanum ssp. triviale

Pink family *Caryophyllaceae*
Flowering time: April–Aug.
Height of growth: 10–30 cm
Dicotyledonous; Perennial

Identification marks
Plant covered with short hairs, and somewhat sticky and glandular. Stem base gnarled, stem erect; barren stems prostrate. Leaves opposite, linear-lanceolate, small-leaved short shoots rooted at the nodes. Terminal cymes. Flowers large; petals about 1 cm long, much longer than the sepals, bifid at tip.

Habitat and distribution
Grassland, field margins, path verges, dry banks. Likes calcareous sandy soils low in nitrogen and especially warm. Scattered throughout Britain, mainly in the east.

Additional information
Similar: Snow-in-summer (*C. tomentosum*). White woolly-haired. Rockery plant, often growing wild.

Pink family *Caryophyllaceae*
Flowering time: April–Sept.
Height of growth: 10–45 cm
Dicotyledonous; Perennial

Identification marks
Stem prostrate or erect, densely hairy, rarely glabrous. Leaves narrow-ovoid, opposite. Petals deeply bifid, as long as calyx (around 0.5 cm). Plant usually dark green.

Habitat and distribution
Meadows, pastures, waysides, dunes and shingle, less common on fields. Widespread; up to around 1,200 m in Scotland. Usually on well-moistened (sandy) loamy soil, rich in nutrients.

Additional information
Earlier names: *C. caespitosum*, *C. triviale*, *C. holosteoides*. Hard to distinguish other species: Sticky Mouse-ear (*C. glomeratum*), arable fields, Little Mouse-ear (*C. semidecandrum*) and Grey Mouse-ear (*C. brachypetalum*) in dry turf.

Corn Spurrey
Spergula arvensis

Three-nerved Sandwort
Moehringia trinervia

Pink family *Caryophyllaceae*
Flowering time: June–Aug.
Height of growth: 7–40 cm
Dicotyledonous; Annual

Identification marks
Stem ascending to erect, round, swollen nodes. The leaves are opposite and appear verticillate through short shoots growing out of leaf axils. Linear to awl-like, longitudinal furrow beneath. Forked flower clusters, petals entire, rarely with pink tinge.

Habitat and distribution
Root-crop fields, rubbish tips, verges. Scattered, sometimes abundant. Sand indicator; likes acid soil which is not too dry, rich in nutrients, low in lime.

Additional information
Some types which grow up to 1 m high are cultivated as fodder plants.
Similar: *S. pentandra*; leaves without furrow; heaths, sandy areas; rare introduction in Sussex.

Pink family *Caryophyllaceae*
Flowering time: May–June
Height of growth: 10–40 cm
Dicotyledonous; Annual

Identification marks
Stem prostrate, ascending or erect, round, covered with downy hairs. Leaves ovoid, pointed, usually with 3 parallel veins (more rarely 5); opposite, lower ones petiolate. Flowers petiolate, emerging from leaf axils; petals shorter than calyx, sometimes only 4 petals.

Habitat and distribution
All types of woodland but preferably deciduous. Likes moist, loamy soil, superficially acid and rich in nutrients; calcifugous. Frequent.

Additional information
Similar: Chickweed (*Stellaria media*, see p.50) with its forest versions. Can be distinguished by its single line of stem hairs.

White Campion
Silene alba

Nottingham Catchfly
Silene nutans

Pink family *Caryophyllaceae*
Flowering time: May–Sept.
Height of growth: 30–100 cm
Dicotyledonous; Biennial – Perennial

Identification marks
Stem erect, forked branching. Leaves opposite, ovoid. Entire plant short-haired to glandular-downy. Three flowers in each leaf axil. Calyx cylindrical-inflated. Petals deeply cleft into two. Flowers unisexual.

Habitat and distribution
Weed communities in fields, wasteland, thickets and hedgerows. Likes warmth and nitrogen. Common. Up to 400 m in Scotland.

Additional information
Synonym: *Melandrium album*. Similar: Night-flowering Catchfly (*S. noctiflorum*), flowers hermaphrodite, petals yellow beneath, pink above, inrolled during the day, only 3 (not 5) styles. Local in sandy arable fields.

Pink family *Caryophyllaceae*
Flowering time: May–July
Height of growth: 25–80 cm
Dicotyledonous; Perennial

Identification marks
Stem erect, soft hairs towards the base, glandular-sticky at the top. Leaves opposite, narrow lanceolate, lower ones spatulate, petiolate. Panicles drooping. Calyx gamosepalous, cylindrical to funnel-shaped, with 10 longitudinal veins. Petals deeply cleft. Only spread out in the evening.

Habitat and distribution
Dry places, rocks, cliffs, edges of fields. Local. Usually on poor, dry-warm, porous and often stony soil. Woodland in N. Europe.

Additional information
Similar: Forked Catchfly (*S. dichotoma*), inflorescence clearly two-forked. Frequent as a casual in fields, on balks, rubbish tips and railway embankments.

Bladder Campion
Silene vulgaris

Soapwort
Saponaria officinalis

Pink family *Caryophyllaceae*
Flowering time: June–Aug.
Height of growth: 25–90 cm
Dicotyledonous; Perennial

Identification marks
Glaucous, usually glabrous plant. Stem erect to ascending. Leaves opposite, ovoid to lanceolate, pointed. Loose cymes. Calyx gamosepalous, pale, inflated, with a network of 20 longitudinal veins. Main veins usually reddish or blue-green. Petals deeply cleft.
Habitat and distribution
Fields, waysides, rocks, dry turf and thickets. Likes warmth, mainly on calcareous soils rich in nutrients. Common throughout most of Britain, less in the north.
Additional information
Synonyms: *S. inflata, S. cucubalus.*

Pink family *Caryophyllaceae*
Flowering time: July–Sept.
Height of growth: 30–90 cm
Dicotyledonous;Perennial

Identification marks
Leaves opposite, broadly ovoid on erect stem which is often tinged with red. Flowers in dense clusters on main stem and its branches. Petals flat, extending from long cylindrical calyx, slightly notched at tip, each with 2 small teeth in the throat of the corolla.
Habitat and distribution
Frequent. A plant of hedgerows and waysides, typically near habitation where it has probably escaped from cultivation. Genuinely wild plants occur next to river banks and in weedy places on porous soil moistened by ground water, further away from bank.
Additional information
Often white and pink versions in same habitat.

Christmas Rose
Helleborus niger

Narcissus-flowered Anemone
Anemone narcissiflora

Buttercup family *Ranunculaceae*
Flowering time: Jan–April
Height of growth: 10–30 cm
Dicotyledonous; Perennial

Identification marks
Flowers single; strong round stem with 1–3 scale-like bracteoles. Basal leaves leathery, evergreen, palmate; long petioles. Flowers large, approx. 5–10 cm across, after end of flowering period turn green or reddish.

Habitat and distribution
Not British. In mixed woodland or pure coniferous forests (pine) in mountains. Likes stony ground rich in humus with sufficient lime and nutrients. Very rare. Grows wild only in the eastern Alps in Germany; rarely grows wild from gardens and parks.

Additional information
In gardens the Southern Alpine type 'Snow Rose' is often grown which has flowers around 10 cm in width.

Buttercup family *Ranunculaceae*
Flowering time: May–Aug.
Height of growth: 20–40 cm
Dicotyledonous; Perennial

Identification marks
Stem erect, round, hairy. Whorl of narrowly divided stem leaves towards the tip, subtending 3–8 pedicellate flowers. Perianth segments 5–60. Basal leaves long-stalked, divided into 3–5 lobes, margins narrow-toothed; dense hairs.

Habitat and distribution
Not British. Semi-dry turf, Alpine tussocks and bushy slopes. Very rare. Only in Schwäbisches Alb and in the Alps. Calcicolous; on well-moistened soils between 700 and 2,400 m.

Additional information
On rocky slopes in the High Alps the variation *dubia* (= *oligantha*) can be found. This has a single flower, is small and glabrous and resembles Wood Anemone (p.85).

Snowdrop Windflower
Anemone sylvestris

Aconite-leaved Buttercup
Ranunculus aconitifolius

Buttercup family *Ranunculaceae*
Flowering time: April–June
Height of growth: 5–40 cm
Dicotyledonous; Perennial

Identification marks
Entire plant covered with hairs. Stem erect, above the middle is a whorl of 3 palmately lobed stem leaves. Basal leaves long-stalked, palmately lobed. Flower up to 7 cm wide, in the centre a cluster of oval-globular pistils, surrounded by a thick crown of anthers. Fruit white, woolly-haired.

Habitat and distribution
Not British. Open woodland, hedgerows, semi-dry turf. Very rare. Likes dry-warm soils containing lime.

Additional information
Similar: Alpine Pasqueflower (*Pulsatilla alpina*, p.86). Mostly over 5 petals, fruit has tail. Rare under 1,000 m.

Buttercup family *Ranunculaceae*
Flowering time: May–July
Height of growth: 20–120 cm
Dicotyledonous; Perennial

Identification marks
Stem erect, branched. Basal leaves long-stalked. Stem leaves alternate and sessile. Leaves divided into 3–7 lobes. Many flowers. 1–2 cm across, pedicellate.

Habitat and distribution
Not British. Wet meadows, banks of streams, marshland round springs, arable land, thickets of high bushes and open canyon forests. Likes damp, nitrogenous (fertilized) slightly acid soils in cool rainy areas. Scattered in mountainous regions up to about 200 m.

Additional information
Similar (usually only as subspecies): Large White Buttercup (*R. platanifolius*). Leaf lobes similar, central lobe not stalked; flower stems glabrous. More common in shady places (woodland, thicket).

River Water-crowfoot
Ranunculus fluitans

Common Water-crowfoot
Ranunculus aquatilis

Buttercup family *Ranunculaceae*
Flowering time: June–Aug.
Height of growth: 1–6 m
Dicotyledonous; Perennial

Identification marks
Only in running water. Stem low floating, usually without floating leaves. Submerged leaves alternate, repeatedly subdivided into thin wisps; 7–30 cm long, stretched out parallel. Single flowers, on long stalks above the water, 2–3 cm in diameter.

Habitat and distribution
Often en masse in clean flowing water containing much oxygen, up to 3 m water depth. Scattered throughout Britain except far north. Likes water to be none too warm.

Additional information
Often placed in the subgenus *Batrachium* with other white-flowering species of the Buttercup family.

Buttercup family *Ranunculaceae*
Flowering time: May–June
Height of growth: 2.5–120 cm
Dicotyledonous; Annual – Perennial

Identification marks
Usually has floating leaves. Submerged leaves alternate, divided into many short wisps. Single flowers, long-stalked, often 1–3 cm across.

Habitat and distribution
In stagnant or only slow-flowing water, rich in nutrients (also slightly polluted) up to 1 m or so water depth. Calcifugous. Scattered throughout lowland regions; rarely on land.

Additional information
Very many variations; also closely related species. Thread-leaved Water-crowfoot (*R. trichophyllus*), flowers less than 1.5 cm wide, usually no floating leaves, repeatedly divided into groups of 3 segments. Similar: Fan-leaved Water-crowfoot (*R. circinatus*), calci-colous, leaves all submerged spread out into flat semi-circle or wheel shape.

Round-leaved Sundew
Drosera rotundifolia

Great Sundew
Drosera anglica

Sundew family *Droseraceae*
Flowering time: June–Aug.
Height of growth: 6–25 cm
Dicotyledonous; Perennial

Identification marks
Stem erect, without leaves. Comes out from centre of a rosette of basal leaves and is at least twice as long as these. Leaves have long stalks, blades are rounded, with reddish 'tentacles' (hairs with sticky globular heads). Few flowers, in coiled cymes.

Habitat and distribution
On high moors, less common in fens and lower moorland. Restricted to suitable peaty areas, but when it occurs, it often does so in large numbers.

Additional information
Together with the following species it forms the hybrid *D. obovata*, recognizable by its ovoid leaf blades and the central inflorescence stalks which extend far beyond the rosette of leaves. Sterile. Very rare.

Sundew family *Droseraceae*
Flowering time: July–Aug.
Height of growth: 10–30 cm
Dicotyledonous; Perennial

Identification marks
Stem erect, without leaves. Comes out from the centre of a rosette of basal leaves and is up to twice as long as these. Leaves petiolate, leaf blades linear-oblong, with reddish tentacles (hairs with sticky globular heads). Few flowers, in coiled cymes.

Habitat and distribution
Moist places on heaths, moors, and fenland. Scattered throughout Britain.

Additional information
Similar: Oblong-leaved Sundew (*D. intermedia*), inflorescence only slightly longer than the leaves, springs from outside the rosette and bends upwards. Drier areas of heath and moorland than other species. Local with a westerly distribution. Compare also *D. obovata* – see left.

Grass-of-Parnassus
Parnassia palustris

Meadow Saxifrage
Saxifraga granulata

Saxifrage family *Saxifragaceae*
Flowering time: July–Oct.
Height of growth: 10–30 cm
Dicotyledonous; Perennial

Identification marks
Stem angular, erect with terminal inflorescence and one cordiform leaf on the bottom third of the stem and clasping it. Basal leaves petiolate, heart-shaped. Entire plant glabrous. Flowers 1–3 cm wide, with 5 normal anthers and 5 long glandular fringed formations (staminodes – transmuted stamens; to attract insects).

Habitat and distribution
Marshland and wet flushes in moorland, in mountainous areas also on moistened detritus. Calcicolous. Widespread, but local though it often occurs in great abundance.

Additional information
A shorter, leathery-leaved variety occurs in dune slacks, in north-west Britain.

Saxifrage family *Saxifragaceae*
Flowering time: April–June
Height of growth: 10–50 cm
Dicotyledonous; Perennial

Identification marks
Stem erect, few leaves, covered with sticky hairs, bulbils present at base of stem (serve for plant reproduction). Basal leaves long-stalked, round to reniform, lobate-crenate, stem leaves 3–5 lobes. Inflorescence a loose cyme with few flowers. Petals approx. 1.5 cm.

Habitat and distribution
Semi-dry turf, meadows not too damp, grassy slopes, more rarely in light woodland and forest margins. Calcifugous; prefers slightly moist soils rich in nutrients, in lower-lying land. Not usually over 425 m. Local, with easterly distribution.

Livelong Saxifrage
Saxifraga paniculata

Stone Bramble,
Stone Blackberry
Rubus saxatilis

Saxifrage family *Saxifragaceae*
Flowering time: May–Aug.
Height of growth: 10–40 cm
Dicotyledonous; Perennial

Identification marks
Stem stiff and erect, branching only at the top, with small alternate leaves. Basal leaves in densely tufted rosette, stiff, rather thick, glaucous, finely toothed and lime-encrusted on the edges. Flowers in clusters, often red-spotted.
Habitat and distribution
Not British. Rocks, walls, scree and stony ground. Prefers dry-warm locations. Calcicolous.
Additional information
Older name: *S. aizoon.*

Rose family *Rosaceae*
Flowering time: June–Aug.
Height of growth: 8–40 cm
Dicotyledonous; Annual

Identification marks
Flowering stem erect, hardly any branches, usually covered with weak prickles. Sterile stems coiled and prostrate to far-creeping with branches which root at the tips. Leaves alternate, petiolate, ternate; leaflets ovoid, margin toothed. Few flowers, in a terminal cyme. Fruit red, bramble-like, but very loose and meagre.
Habitat and distribution
In all types of woodland on calcareous soils which are not too dry and contain abundant humus. Local, more common in hilly areas.
Additional information
Similar: low or creeping versions of the Bramble (*R. fructicosus*, p.112), stem woody.

Hautbois Strawberry
Fragaria moschata

Wild Strawberry
Fragaria vesca

Rose family *Rosaceae*
Flowering time: April–July
Height of growth: 10–40 cm
Dicotyledonous; Perennial

Identification marks
Stem erect, clinging hairs at the top. Leaves ternate, leaflets ovoid, serrate. Cyme with few flowers. Petals pale ivory in colour, margin entire, touching each other. Plant devoid of runners. Fruit (strawberry) greenish to purplish-red and has clinging sepals (hard to remove).

Habitat and distribution
Formerly widely cultivated. Reported as an escape in a number of areas but many claims probably erroneous.

Additional information
Similar: other species of strawberry and Barren Strawberry (see right). All strawberry species easily form hybrids with one another.

Rose family *Rosaceae*
Flowering time: April–July
Height of growth: 5–30 cm
Dicotyledonous; Perennial

Identification marks
Flowering stem erect, clinging hairs at the top; runners arching, spreading from the base. Leaves ternate. Leaflets ovoid, serrate margin. Cyme with 3–10 flowers. Petals touch or overlap, with entire margins. Red fruit.

Habitat and distribution
All types of woodland, woodland paths, clearings and pastures. Likes rather moist, base-rich soil which is not too shady. Common.

Additional information
Similar: Hautbois Strawberry (adjacent) and Barren Strawberry (see next page).

Barren Strawberry
Potentilla sterilis

White Cinquefoil
Potentilla alba

Rose family *Rosaceae*
Flowering time: Feb.–May
Height of growth: 5–15 cm
Dicotyledonous; Perennial

Identification marks
Flowering stem, prostrate to ascending, 1–3 flowers, protruding hairs. Leaves ternate, overall similar to strawberry, often bluish-green, hairy. Flowers 1–1.5 cm wide, their petals hardly extend beyond calyx, margin not quite entire at tip, not touching each other.

Habitat and distribution
Woodland, thickets, balks, mossy grass pastures. Calcifugous, in drier soils. Common, rarer in the north, reaching to over 700 m.

Additional information
Synonym: *P. fragariastrum*. Very similar: Wild Strawberry (see left-hand page), note distinguishing features of petals.

Rose family *Rosaceae*
Flowering time: April–June
Height of growth: 5–20 cm
Dicotyledonous; Perennial

Identification marks
Stem decumbent or more or less erect, few small leaves, usually 3 flowers to each stem. Basal leaves long-stalked, digitate with 5 lobes; lobes lanceolate, serrate at tip, underside covered with dense silver-white hairs. Flowers long-stalked, approx. 2 cm wide.

Habitat and distribution
Not British. Open dry woodland, sunny thickets. Rare. Likes warmth.

Goat's-beard Spiraea
Aruncus dioicus

Meadowsweet
Filipendula ulmaria

Rose family *Rosaceae*
Flowering time: April–July
Height of growth: 80–150 cm
Dicotyledonous; Perennial

Identification marks
Stem stiff erect, glabrous, simple. Leaves up to 0.5 cm long, bi- or tripinnate, when young usually copper-red in colour. Leaflets ovoid, pointed at the tip, double serrate. Panicle up to 50 cm long, many-flowered, usually composes only one sex of flower: male flowers ivory-coloured, around 4 mm wide; female flowers milk-white, around 3 mm wide.
Habitat and distribution
Not British. Canyon forests, mountain forests, shady banks of streams. Scattered. Likes soils rich in nutrients but low in lime and moist through seepage.
Additional information
Synonyms: *A. vulgaris*, *A. sylvestris*.

Rose family *Rosaceae*
Flowering time: June–Sept.
Height of growth: 60–120 cm
Dicotyledonous; Perennial

Identification marks
Stem erect, angular, glabrous, often branched at the top. Main stem leaves are numerous, simple pinnate, with 2–5 pairs of large leaflets interrupted by further pairs of much smaller leaflets. Cymose panicle with numerous strongly scented flowers, up to 1 cm wide. Flowers with 6 petals sometimes occur.
Habitat and distribution
Marshy meadows, ditches, banks of streams, wet thickets and water meadows. Prefers wet soil rich in nutrients. Common; up to about 1,000 m.
Additional information
Similar: Dropwort (*F. vulgaris*), 8–20 pairs of main leaflets; flowers all white with 6 petals and 6 sepals. Chalky grassland. Widespread but local.

Wood Sorrel
Oxalis acetosella

Purging Flax
Linum catharticum

Wood-sorrel family *Oxalidaceae*
Flowering time: April–May
Height of growth: 5–15 cm
Dicotyledonous; Perennial

Identification marks
All flowers and leaves long-stalked, basal. Leaves clover-like, ternate with heart-shaped leaflets which are often reddish on the underside. Lower part of peduncle has a pair of scale-like subtending leaves. Petals about 1–1.5 cm long, several times longer than the calyx, fine violet or red veins, with yellow fleck at the base.

Habitat and distribution
In hedgerows and all types of woodland on well-moistened, slightly acid and porous soil which is not too heavy or wet. Likes shade. Very common.

Flax family *Linaceae*
Flowering time: June–Sept
Height of growth: 5–25 cm
Dicotyledonous; Annual – Biennial

Identification marks
Stem erect, thin, little or no branching. Stem leaves opposite, entire, narrow oblong, somewhat rough; the upper leaves occasionally alternate. Loose cymes. Petal margins entire, 4–5 mm long.

Habitat and distribution
Wet meadows, semi-dry turf, pathways, acid grass meadows. Likes poor soils which are wet in winter and dry in summer. Characteristic of calcareous grassland. Common.

Additional information
Could be mistaken for a type of Stitchwort (*Stellaria*, pp.50, 51) or Chickweed (*Cerastium*, p.52), which normally, however, have cleft petals.

Sanicle
Sanicula europea

Field Eryngo
Eryngium campestre

Umbellifer family *Apiaceae (Umbelliferae)*
Flowering time: May–Sept.
Height of growth: 20–60 cm
Dicotyledonous; Perennial

Identification marks
Stem erect, stiff, sometimes twisted, angular. Few small stem leaves. Basal leaves large, coarse, long-stalked, palmate with 3–5 wide segments. These are in turn lobed and serrate. Compound umbels, the individual small umbels are globular. Flowers occasionally have reddish tinge.

Habitat and distribution
Deciduous and mixed woodland. Frequent. Likes wet loamy soil rich in mull. Only found in shady or semi-shady locations. From lowlands to relatively high hilly areas.

Additional information
The plant was previously used for medicinal purposes and bore a variety of popular names.

Umbellifer family *Apiaceae (Umbelliferae)*
Flowering time: July–Aug.
Height of growth: 30–60 cm
Dicotyledonous; Perennial

Identification marks
Stem erect, strong, finely furrowed, usually branching in top third. Leaves coarse, light grey-green, net venation and pinnate, thistle-like, with spiny margins. Lower leaves petiolate, upper ones amplexicaule. Almost spherical umbels with narrow or leaf-like, spiny terminal involucres.

Habitat and distribution
Dry or semi-dry turf. Likes the warmth, calcicolous. Likes dry, often stony ground in full sunlight. Rare, only in few localities in southern England.

Ground Elder
Aegopodium podagraria

Caraway
Carum carvi

Umbellifer family *Apiaceae (Umbelliferae)*
Flowering time: May–July
Height of growth: 40–100 cm
Dicotyledonous; Perennial

Identification marks
Stem erect, some branching, furrowed, hollow, few leaves at the top, basal leaves numerous because of subterranean runners, simple or doubly tripartite with large ovoid, deeply serrate leaflets. Compound flat umbels with approx. 15 rays; bracts and bracteoles absent.

Habitat and distribution
A common and sometimes troublesome weed of waste places and especially in gardens. Likes nitrogenous soil which is moist with ground water. Widely distributed.

Additional information
Previously used for medicinal purposes: Goutweed, Herb Gerard, Bishop's Weed.

Umbellifer family *Apiaceae (Umbelliferae)*
Flowering time: June–July
Height of growth: 25–60 cm
Dicotyledonous; Perennial

Identification marks
Stem erect, much branched, leafy, fluted. Leaves bi- or tripinnate with very narrow tips. Stem leaves inflated at base where the lowest pinnae are situated. Compound umbel with 5–10 rays. Bracts and bracteoles absent (sometimes 1).

Habitat and distribution
In waste places. Prefers soils which are not too dry but rich in nutrients. Scattered and rather rare.

Additional information
Similar plants of the Umbellifer family (see next page), but the distinguishing feature of this one is the lowest pinnae which have 'slipped' down the stem.

Cow Parsley
Anthriscus sylvestris

Hogweed
Heracleum sphondylium

Umbellifer family *Apiaceae (Umbelliferae)*
Flowering time: April–June
Height of growth: 60–100 cm
Dicotyledonous;.Biennial

Identification marks
Stem erect, branched, furrowed, hollow. Leaves bi- or tripinnate; segments coarsely toothed. Compound umbel with 4–10 rays, bract absent, bracteoles ciliate. Petals notched with an inflated tip.

Habitat and distribution
Meadows, meadows with trees, hedges, wayside verges. Prefers slightly wet porous soils with plenty of nutrients. Often in great abundance in meadows fertilized with liquid manure. Very common.

Additional information
Very similar: Golden Chervil (*Chaerophyllum aureum*); stem solid and usually has red or purple flecks. Meadows, in several localities in Scotland.

Umbellifer family *Apiaceae (Umbelliferae)*
Flowering time: June–Sept.
Height of growth: 50–200 cm
Dicotyledonous; Biennial

Identification marks
Stem erect, angular and grooved, thick, covered with stiff hairs. Leaves usually once pinnate, the segments lobed or further divided, with rough hairs, lower leaves petiolate. Compound umbel with 7–20 rays. Bracts absent or few in number. Bracteoles numerous. Marginal flowers with outer petals enlarged.

Habitat and distribution
Meadows, banks, woods, damp thickets, water meadows. Prefers damp soils rich in nutrients. Common. Indicator of over-fertilization.

Wild Carrot
Daucus carota

Wild Angelica
Angelica sylvestris

Umbellifer family *Apiaceae (Umbelliferae)*
Flowering time: June–Aug.
Height of growth: 30–100 cm
Dicotyledonous; Biennial

Identification marks
Stem erect, often branched, grooved, hairy, solid. Leaves tripinnate. Compound umbels, initially resemble bird's nest, becoming flat when flowering and when in fruit they revert to bird's nest shape. Bracts pinnate. Central flower in umbel usually blackish purple ('Blackamoor Flower').

Habitat and distribution
Semi-dry turf, meadows, wayside verges, cliffs and dunes. Frequent. Likes the warmth.

Additional information
Similar: Cambridge Milk-parsley (*Selinum carvifolia*), bract is usually absent; same applies to Greater Burnet-saxifrage and Burnet-saxifrage (*Pimpinella major* and *P. saxifraga*), where the stem leaves are simple pinnate.

Umbellifer family *Apiaceae (Umbelliferae)*
Flowering time: July–Sept.
Height of growth: 30–200 cm
Dicotyledonous; Perennial

Identification marks
Compound umbel with 20–40 rays. Numerous bracteoles, bracts absent or reduced to a few and falling early. Stem strong, hollow, has whitish tinge, basic colour bluish to reddish. Leaves 2–3 pinnate; segments ovoid, serrate margin. Leaf sheaths striking, light-coloured and inflated.

Habitat and distribution
Water meadows, banks, damp meadows or thickets and woods in places on soils rich in nutrients near to ground water. Common.

Additional information
Wild Angelica is one of those plants in the Umbellifer family which have either white or pink flowers (see p.229).

Marsh Pennywort
Hydrocotyle vulgaris

One-flowered Wintergreen
Moneses uniflora

Umbellifer family *Apiaceae (Umbelliferae)*
Flowering time: June–Aug.
Height of growth: 1–25 cm
Dicotyledonous; Perennial

Identification marks
Leaves round, notched along margin. Petiole comes out of middle of leaf underside. Umbels are small, few flowers, from leaf axils. Peduncle decidedly shorter than petiole. Stem long and creeping, 10–50, sometimes 100, cm long.

Habitat and distribution
Bogs, marshes, ditches, banks. Likes soils low in lime and moistened by ground water. Frequent, occurring up to about 600 m.

Wintergreen family *Pyrolaceae*
Flowering time: June–Aug.
Height of growth: 1–5 cm
Dicotyledonous; Perennial

Identification marks
Solitary flower on leafless stem, drooping. Petals spread out flat; flower around 1.5 cm in diameter, sweet-scented. Leaves in rosette arrangement, evergreen, leathery, rounded, the blade extending somewhat down the short stalks. Exceptionally 2–3 flowers on one stem.

Habitat and distribution
Coniferous woodland. Rare. Likes mossy, somewhat acid soil, dry to slightly moist. Restricted to E. Scotland.

Additional information
Listed under Wintergreens as *Pyrola (=Pirola) uniflora*.

Serrated Wintergreen
Orthilia secunda

Round-leaved Wintergreen
Pyrola rotundifolia

Wintergreen family *Pyrolaceae*
Flowering time: July–Aug.
Height of growth: 2–10 cm
Dicotyledonous; Perennial

Identification marks
Drooping, campanulate flowers in unilateral racemes. Anthers protrude a little beyond the flower, style noticeably so. Stem ascending. Leaves evergreen, leathery-tough, ovoid, pointed, margin finely serrate. Initially the entire cluster of 20–30 flowers droops.

Habitat and distribution
Mixed and coniferous woodland. Rather local in higher areas of N. England, Scotland and Wales as high as 800 m.

Additional information
This plant also included in other genera as: *Ramischia secunda, Pyrola secunda* (= *Pirola secunda*).

Wintergreen family *Pyrolaceae*
Flowering time: July–Sept.
Height of growth: 20–35 cm
Dicotyledonous; Perennial

Identification marks
Drooping wide campanulate flowers in multilateral raceme. Only the style projects beyond the flower. Stem erect, obtuse-angled. Leaves almost round, stalked, tough-leathery, finely crenate, in basal rosette. Local with strong easterly distribution, being absent from many parts of the west.

Habitat and distribution
Mixed and coniferous woodland. Likes moist somewhat acid loamy soil which is low in lime and rich in humus.

Additional information
Similar species in the same genus of *Pyrola* (= *Pirola*), all rare and growing in woodland: Intermediate Winter-green (*P. media*), flowers globular (-campanulate), style projecting; Common Wintergreen (*P. minor*), flowers globular, style hidden.

Milkwhite Rock-jasmine
Androsace lactea

Water-violet
Hottonia palustris

Primrose family *Primulaceae*
Flowering time: May–July
Height of growth: 5–15 cm
Dicotyledonous; Perennial

Identification marks
Dense rosette of linear-lanceolate leaves. Out of the middle of leafless erect flower stalk with loose terminal umbel with few flowers. Flowers approx. 1 cm wide, with yellow throat and notched petals. Rosettes usually numerous in porous turf.
Habitat and distribution
Not British. Limestone fissures in the Alps up to 2,200 m. Scattered. Also found on the Schwäbisches Alb.

Primrose family *Primulaceae*
Flowering time: May–June
Height of growth: 15–40 cm
Dicotyledonous; Perennial

Identification marks
Leaves in whorls/rosettes, pinnate comb-like, usually submerged. Flower stem above water, leafless, erect. Whorls of flowers forming cluster. Flowers approx. 2 cm across with yellow throat, short-stalked, ascending in flower, deflexed in fruit.
Habitat and distribution
Stagnant or slow-flowing waters, ponds, ditches. Widespread but local in England, Wales and a few parts of Scotland. Most common in E. England. Prefers soils which are rather poor in nutrients and low in lime and above all likes shallow water.
Additional information
Occasionally the flowers have a somewhat reddish tinge, especially with the rare flowering deep water types (1–2 m deep), where they are submerged.

Bogbean
Menyanthes trifoliata

Vincetoxicum
Vincetoxicum hirundinaria

Bogbean family *Menyanthaceae*
Flowering time: May–July
Height of growth: 12–30 cm
Dicotyledonous; Perennial

Identification marks
Leaves divided like Clover, thickish-leathery, on long stalks from a creeping rhizome. Leaflets ovoid, up to 7 cm long. Stem ascending. Dense raceme. Petals have conspicuous hairs on the inner surface, bud usually has red tinge.

Habitat and distribution
Bogs, fens and moorland, banks, ditches, ponds; often spreads far out into shallow water. Quite common, sometimes very abundant. Calcifugous; grows up to about 1,000 m.

Additional information
Was an old medicinal plant.

Milkweed family *Asclepiadaceae*
Flowering time: May–Aug.
Height of growth: 30–120 cm
Dicotyledonous; Perennial

Identification marks
Stem erect, sometimes twining towards the top (see below). Leaves opposite, ovate-cordate, shortly petiolate. Flowers in axillary racemes. Fruit pod-like, with many seeds with hairy heads.

Habitat and distribution
Not British. Dry woodland, sunny thickets, balks, rocks and screes. Prefers calcareous stony ground rich in nutrients. Scattered.

Additional information
The twining version (*var. laxum*) has yellowish-white flowers with green tips. That is the mountain version found in the south of Germany. Synonyms for the entire species are: *V. officinale* and *Cynanchum vincetoxicum*.

Hedge Bindweed
Calystegia sepium

Field Gromwell
Buglossoides arvensis

Convolvulus family *Convolvulaceae*
Flowering time: July–Sept.
Height of growth: 1–3 m
Dicotyledonous; Perennial

Identification marks
Climbing stem. Leaves cordate to sagittate, petiolate, alternate. Flowers funnel-shaped, up to 7 cm long; stems coil in anti-clockwise direction.

Habitat and distribution
Hedgerows, waste places, rubbish tips, fences. Frequent on rich soils in mild locations.

Additional information
Similar species: Large Bindweed (*C. silvatica*), corolla white, about 7 cm long; Hairy Bindweed (*C. pulchra*), corolla about 6 cm, bright pink with light stripes; Sea Bindweed (*C. soldanella*), leaves reniform, flowers pink with 5 white stripes. Found only on dunes, sea shores, rarer in N. Scotland.

Borage family *Boraginaceae*
Flowering time: April–June.
Height of growth: 10–50 cm
Dicotyledonous; Annual – Biennial

Identification marks
Plant coarsely hairy. Stem erect, simple or branched. Leaves alternate, lanceolate, with one vein. Flowers small, barely 5 mm wide.
Habitat and distribution
Not British. Cornfields, verges of ploughed fields. On sand and loamy soil which is rich in nutrients but not too dry. Scattered, rarely over 1,000 m.

Additional information
Contains a red dye which was previously used as a rouge. Was known for a long time as *Lithospermum arvense*. Similar: Common Gromwell (*Lithospermum officinale*), flowers more of a dirty or greenish white, leaves have conspicuous lateral veins, plant 0.3–1 m; sunny ridges, thickets; rare.

White Mullein
Verbascum lychnitis

Black Nightshade
Solanum nigrum

Figwort family Scrophulariaceae
Flowering time: June–Aug.
Height of growth: 50–150 cm
Dicotyledonous; Biennial

Identification marks
Entire plant covered with fine hairs ('downy').
Stem erect, often somewhat branched,
robust. Leaves ovoid, dark green above, lower
ones petiolate, upper ones sessile, alternate.
Long dense panicle comprising numerous
clusters of 2–5 flowers. Flowers 1–2 cm wide,
flat. Stamens covered with thick white wool.

Habitat and distribution
Dry sunny banks, railway embankments,
wayside verges, and waste places. Calcicolous
nitrogen indicator. Local in S. England and
Wales, casual further north.

Additional information
Occasionally versions crop up with virtually
yellow flowers.

Nightshade family Solanaceae
Flowering time: July–Sept.
Height of growth: 30–60 cm
Dicotyledonous; Annual

Identification marks
Stem branched, prostrate to erect, usually
glabrous. Leaves ovoid to rhomboid, entire or
slightly lobed margin, alternate, stalked.
Corolla at first spread out flat then curving
backwards, anthers large and inclining
towards each other in a cone-shape. Fruit is a
berry, green when immature, ripening black.

Habitat and distribution
Weedy places in gardens, fields, waste
ground, rubbish tips and walls. On loamy
soils with abundant nutrients. Common, local
in Wales, becoming rarer northwards. Almost
absent from Scotland.

Dwarf Elder
Sambucus ebulus

Three-leaved Valerian
Valeriana tripteris

Honeysuckle family *Caprifoliaceae*
Flowering time: July–Aug.
Height of growth: 60–120 cm
Dicotyledonous; Perennial

Identification marks
Stem stiff and erect, little branching, stout but thoroughly herblike. Leaves opposite, simple or bipinnate. Leaflets of pinnate leaves ovoid, serrate margin. Terminal many-flowered racemes with 3 main branches. Fruits are small, black, globular berries.
Habitat and distribution
In waste places and along roadsides. Needs moist soils rich in nutrients. Calcicolous. Very scattered, local in most parts, more so in Scotland.
Additional information
Also known as Danewort. Similar: Elder (*S. nigra*, see p.115), usually branched but above all woody.

Valerian family *Valerianaceae*
Flowering time: April–June
Height of growth: 10–50 cm
Dicotyledonous; Perennial

Identification marks
Leaves matt, the upper ones three-lobed, the lower ones entire, heart-shaped, toothed. Stem erect, unbranched. Many-flowered paniculate inflorescence.
Habitat and distribution
Not British. Rock fissures, scree, stony woodland and thickets in mountainous areas. Calcicolous, needs moisture; likes light and semi-shade. Only found in the area of the Alps and outlying regions (Alb, Black Forest, plateau). Scattered. Up to about 2,000 m.

White Bryony
Bryonia cretica ssp. *dioica*

Spiked Rampion
Phyteuma spicatum

Gourd family *Cucurbitaceae*
Flowering time: May–Sept.
Height of growth: 0.5–4 m
Dicotyledonous; Perennial

Identification marks
Stem climbs by means of simple tendrils which arise from the side of the deciduous leaves. Leaves coarsely hairy, 5-edged to 5-lobed. Male and female flowers on separate plants, male ones are almost twice as large (approx. 1–2 cm wide), long peduncles. Berries green or whitish when immature, turning red when fully ripe.
Habitat and distribution
Paths, fences, walls, hedges, woodland margins. Prefers calcareous soils, not too dry but rich in nutrients in warm location. Scattered but locally common in the south, rare in the north. Absent in C. and N. Scotland.

Bellflower family *Campanulaceae*
Flowering time: July–Aug.
Height of growth: 30–80 cm
Dicotyledonous; Perennial

Identification marks
Flowers in cylindrical heads, curled up (like a claw) before blossoming. Plant glabrous. Stem erect, simple. Leaves alternate: basal leaves have long stalks, virtually as wide as they are long, cordate; the upper ones narrow ovoid, sessile.
Habitat and distribution
Deciduous and mixed woods and thickets. Prefers porous soils which are not too dry but rich in nutrients and containing mull. Rare. Only recorded from Sussex.

False Helleborine
Veratrum album

Star-of-Bethlehem
Ornithogalum umbellatum

Lily family *Liliaceae*
Flowering time: June–Aug.
Height of growth: 50–150 cm
Monocotyledonous; Perennial

Identification marks
Thick, stiff, erect stem with alternate leaves, the undersides covered with downy hairs. Many panicles, terminal, many-flowered. Often over 0.5 m long. Flowers 1–1.5cm wide.

Habitat and distribution
Not British. Frequent in Alpine pastures, rarer in fenland and water meadows. Nitrogen indicator, likes lime. Only in the Alps and outlying areas up to the Mittelgebirge.

Additional information
Flowers of Alpine variety, ssp. *album*, are white inside, greenish outside. Plus, rare, but in outlying regions of Alps common, ssp. *lobelianum*, flowers more or less greenish. Very similar but does not flower: Great Yellow Gentian (*G. lutea*, p.152) with opposite smooth leaves; also not British.

Lily family *Liliaceae*
Flowering time: April–June
Height of growth: 10–30 cm
Monocotyledonous; Perennial

Identification marks
Loose raceme on stiff stem. Pedicel up to 10 cm long. Petals approx. 2 cm long and at least 4 (to 8) mm wide with green stripe down the back. The grass-like basal leaves with the white central stripe appear only in autumn.

Habitat and distribution
Grassy areas. Local in England, more so in Wales and Scotland. Prefers porous loamy soil rich in nutrients in light or semi-shady locations.

Additional information
Similar: Drooping Star-of Bethlehem (*O. mitans*). Taller, inflorescence unilateral, flowers drooping. Very local in E., C. and N. England.

Branched St. Bernard's Lily
Anthericum ramosum

St. Bernard's Lily
Anthericum liliago

Lily family *Liliaceae*
Flowering time: June–Aug..
Height of growth: 30–80 cm
Monocotyledonous; Perennial

Identification marks
Stem more or less erect, often unbranched and leafless up to the inflorescence. Basal leaves grasslike. Inflorescence is a panicle (made up of several clusters) at the base of which there is often a small deciduous bract. Flowers up to 1.3 cm long, spread out wide and funnel-shaped, with long straight style.
Habitat and distribution
Not British. Dry turf, light woodland and thickets. On sunny calcareous soils. Rare, but usually occurs in abundance. Up to over 1,500 m.
Additional information
Sometimes (stunted) varieties occur with cluster-like inflorescence.

Lily family *Liliaceae*
Flowering time: May–June.
Height of growth: 30–70 cm
Monocotyledonous; Perennial

Identification marks
Stem erect to ascending, usually unbranched and rarely bears leaves. Basal leaves grass-like, inflorescence a simple raceme. Flowers 1.5–2 cm long, funnel-shaped. Style bent, twisted upwards, no longer than the petals.
Habitat and distribution
Not British. Dry turf, light woodland and thickets. Usually on sunny warm ground, low in lime. Rare; absent in the Alps, otherwise up to over 1,200 m.
Additional information
Very rare in habitats rich in minerals and when it grows in well-fertilized gardens the lower part of the inflorescence has short branches.

Ramsons
Allium ursinum

Streptopus amplexifolius

Lily family *Liliaceae*
Flowering time: April–June
Height of growth: 10–45 cm
Monocotyledonous; Perennial

Identification marks
When the plant is rubbed it gives off a garlic smell which is also recognizable even at a distance. Only 2 (or sometimes 3) leaves, ovate-lanceolate and with short stalk. Many-flowered false umbels with obtuse triangular stem, before flowering is enclosed in pale-coloured sheath (the spathe) shaped like narrow onions.

Habitat and distribution
Woods and shady places. Likes loose, moist soil containing plenty of humus and nutrients. Often forms dense and diffuse stands.

Lily family *Liliaceae*
Flowering time: May– July.
Height of growth: 40–100 cm
Monocotyledonous; Perennial

Identification marks
Stem slightly zig-zag in shape, erect. Leaves alternate, oval with heart-shaped base clasping the stem. Flowers deeply divided into 6 segments, solitary in leaf axils, drooping under the respective leaf with their bent and jointed pedicels. Fruit: red berries.
Habitat and distribution
Not British. Mountain forests and heaths, Alpine thickets. Grows on moist, shady soils containing acid humus. Frequent in the Alps, rare in the Mittelgebirge; hardly found under 700 m.
Additional information
Similar: Solomon's-seal (*Polygonatum multiflorum*, see next page), with straight, often hanging stem and leaves narrower at base.

Solomon's-seal
Polygonatum multiflorum

Whorled Solomon's-seal
Polygonatum verticillatum

Lily family *Liliaceae*
Flowering time: May–June
Height of growth: 30–80 cm
Monocotyledonous; Perennial

Identification marks
Stem round, usually hanging a little. Leaves alternate, elliptical, sessile, narrower at base. Often resemble wings stretched out to both sides and pointing slightly upwards. Axillary drooping racemes with 2–5 narrow, funnel-like flowers. Berries globular, blue-black.

Habitat and distribution
In woods, especially under deciduous trees. Local in England and Wales, perhaps naturalized in Scotland. Prefers loose, calcareous loamy soil, rich in humus.

Additional information
Similar: Angular Solomon's-seal (*Polygonatum odoratum = officinale*). Stem angular, flowers usually solitary. Sunny woodland. Very local in N. and W. England and Wales.

Lily family *Liliaceae*
Flowering time: June–July
Height of growth: 30–80 cm
Monocotyledonous; Perennial

Identification marks
Stem thickish, stiff and erect, angular, glabrous. Leaves linear-lanceolate, in whorls of 3–6. Tankard-shaped hanging flowers in leaf axils. Berries first red, when ripe black-blue, globular.

Habitat and distribution
Mixed woods in high locations. Very rare, in N. England and S. Scotland only.

Fritillary
Fritillaria meleagris

Lily-of-the-valley
Convallaria majalis

Lily family *Liliaceae*
Flowering time: April–May
Height of growth: 20–50 cm
Monocotyledonous; Perennial

Identification marks
Stem erect, with 3–6 very narrow deciduous leaves; these are grooved and blue-green. Flowers solitary (rarely in pairs), drooping, campanulate, up to 4 cm long and 2 cm wide, white and purple chequered like a chessboard, rarely creamy.

Habitat and distribution
Very local, in low-lying water meadows rich in nutrients, often flooded in spring. S., E. and C. England.

Additional information
Often cultivated in gardens and then sometimes escaping. Formerly quite widespread but now disappearing.

Lily family *Liliaceae*
Flowering time: May–June
Height of growth: 8–20 cm
Monocotyledonous; Perennial

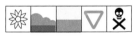

Identification marks
Usually 2 (more rarely 1 or 3) short-stemmed basal leaves, ovate-lanceolate, fine bow-shaped veins, long sheath. Flowering stem little longer than the leaves, erect, with unilateral terminal raceme with few flowers. Flowers drooping, bell-shaped, with 6 tips curved outwards. Fruit is a globular red berry.

Habitat and distribution
Dry woodland, especially where there are deciduous trees. Likes calcareous soils rich in humus in warm locations. Local but quite widespread in England, less so in Wales and Scotland. Has an easterly distribution.

Additional information
Popular spring flower often cultivated for the delicate fragrance of the flowers.

Snowdrop
Galanthus nivalis

Spring Snowflake
Leucojum vernum

Daffodil family *Amaryllidaceae*
Flowering time: Jan.–March
Height of growth: 15–25 cm
Monocotyledonous; Perennial

Identification marks
Flowers solitary, drooping; 3 outer petals large, projecting, 3 inner ones smaller, straight, white with a green spot at the base. Each stem with 2 grass-like somewhat fleshy, slightly glaucous leaves. These are not full grown when the plant flowers. Usually other individuals growing very close produce leaves only.

Habitat and distribution
Deciduous and mixed woodland. Requires soil saturated in ground water and rich in nutrients and mull. Local. Often escapes from gardens and grows wild, becoming naturalized in many places.

Daffodil family *Amaryllidaceae*
Flowering time: Feb.–April
Height of growth: 15–20 cm
Monocotyledonous; Perennial

Identification marks
Stem has 1–2 flowers. Flowers drooping, 2–3 cm long; all 6 petals identical, with yellow-green fleck at the tip. Leaves grass-like, bright green, often longer than the peduncle at flowering time.

Habitat and distribution
Damp hedges and scrub. Prefers loamy soil rich in nutrients and mull. Very rare, wild only in two places in W. England but often cultivated and many escape.

Additional information
Similar plant is the related Summer Snowflake (*L. aestivum*), with 3–6 flowers per peduncle; flowers around May. Very local in S. England.

Spring Crocus
Crocus vernus

White Water-lily
Nymphaea alba

Iris family *Iridaceae*
Flowering time: March–April
Height of growth: 5–15 cm
Monocotyledonous; Perennial

Identification marks
Leaves grass-like with white middle stripe appearing with the flowers. Flowers have narrow petals, at least 4 times as long as they are wide. Flowers only a little hairy in their throats.

Habitat and distribution
Meadows and fields. Local in England and a very few places in Scotland and Wales. Likes calcareous loamy soil, rich in nutrients and moist in the spring.

Additional information
The flowers can be white (ssp. *albiflorus*). Various species of spring-flowering Crocus are cultivated and may escape or even persist in old abandoned gardens.

Water-lily family *Nymphaeaceae*
Flowering time: July–Aug.
Height of growth: 0.5–2.5 cm
Dicotyledonous; Perennial

Identification marks
Large round floating leaves with deep heart-shaped indentation on rope-like stalk. Flowers have many petals, wide open, 10–20 cm across. Stamens yellow.

Habitat and distribution
Stagnant or slow-flowing warm waters, containing some nutrients. Frequent in suitable habitats.

Additional information
In cooler (moorland) lakes poor in nutrients the subspecies occidentalis is found, its flowers 5–10 cm wide. Found in N. Scotland, rare. Occasionally red-flowering ornamental varieties are planted out. The Yellow Water-lily (*Nuphar luteum*, p.132) has similar leaves but without the transverse links between the lateral veins.

Wood Anemone
Anemone nemorosa

Spring Pasqueflower
Pulsatilla vernalis

Buttercup family *Ranunculaceae*
Flowering time: March–May
Height of growth: 6–30 cm
Dicotyledonous; Perennial

Identification marks
The flowers emerge individually out of a whorl composed of three, three-lobed leaves about two-thirds of the way up the stem. Flower 1.5–4 cm wide, glabrous or slightly hairy, often tinged with red on the outside – like the parts of the stem. Usually only 1 petiolate basal leaf, palmately lobed, appearing after the flowers are over.

Habitat and distribution
Woods, especially deciduous ones, thickets. In all but the poorest and wettest soils. Very common, often forming very large colonies.

Additional information
Variations in size, degree of hairiness, number and colour of petals (tinged reddish, bluish, greenish) can be found.

Buttercup family *Ranunculaceae*
Flowering time: April–June
Height of growth: 5–35 cm
Dicotyledonous; Perennial

Identification marks
Flowers solitary, more or less erect, open to form a bell, outside usually bluish-violet. The stem bears a sheathed whorl of deeply divided bracts. Flower and stem covered with furry layer of hairs, usually golden yellow, more rarely white. Leaves leathery, evergreen, simple pinnate.

Habitat and distribution
Not British. Poor sandy turf, open coniferous forests, Alpine mats. Very rare. Prefers porous slightly acid soil containing plenty of humus.

Additional information
The lowland variety (var. *vernalis*) is robust and its flowers are approx. 4.5 cm wide. The Alpine variety (var. *alpestris*) is stumpy, its flowers around 5.5 cm wide. In between them is the mountain variety (var. *bidgostiana*).

Yellow Alpine Pasqueflower
Pulsatilla alpina

Buttercup family *Ranunculaceae*
Flowering time: June–Aug.
Height of growth: 10–40 cm
Dicotyledonous; Perennial

Identification marks
The flowers are solitary and emerge from a whorl of bracts which sheathes the otherwise leafless stem. Stem is covered with tufts of hair. Flowers 4–7 cm wide. Basal leaves petiolate, tripinnate, the individual sections being bipinnate and toothed. Small fruits have tufted hairy tail, numerous.
Habitat and distribution
Not British. Virtually only in limestone Alps. On mountain meadows, stony mats, rocks; also in thickets and pine areas; scattered, rarely under 1,500 m.
Additional information
Only found at a few points in the central European mountain chain; ssp. *alba*, smaller, flowers approx. 3–4 cm across (Vosges, Harz, Riesengebirge, Carpathians); virtually dying out.

Mountain Avens
Dryas octopetala

Rose family *Rosaceae*
Flowering time: June–July
Height of growth: 2–8 cm
Dicotyledonous

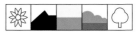

Identification marks
Stem and branches pressed close to the ground, woody. Pedicals erect, approx. 2–8 cm long. Usually 8 petals which wither rapidly. Oblong leaves leathery, evergreen, underside white-woolly, notched along margin and rather curled up. Small fruits with long tufted hairy tails borne in clusters.
Habitat and distribution
On rocks, scree, in crevices. Scattered and local. Usually on mountains but coming down to sea level in N. Scotland. Calcicolous.

Dropwort
Filipendula vulgaris

Rose family *Rosaceae*
Flowering time: May–Aug.
Height of growth: 15–18 cm
Dicotyledonous; Perennial

Identification marks
Stem erect, may be slightly branched at top. Leaves largely basal, simple pinnate; leaflets approx. 2 cm long, serrate, 8–20 pairs of main leaflets per leaf. Many-flowered branched cymose panicle. Flowers usually have 6 petals, more rarely 5.

Habitat and distribution
Meadows and grassland, especially on soils which are wet for a time and then dry again for a while. Calcicolous. Widespread but local, may be quite common in some areas; rarer in Scotland.

Additional information
This is an old medicinal plant. Synonym: *F. hexapetala*.

Chickweed Wintergreen
Trientalis europaea

Primrose family *Primulaceae*
Flowering time: June–July
Height of growth: 10–25 cm
Dicotyledonous; Perennial

Identification marks
Stem erect, with small, alternate leaves at the base. At the top a lax rosette of 5–6 large ovoid-lanceolate leaves. Flowers solitary on long slender pedicels, up to 1.5 cm across. Corolla deeply divided into 7 lobes.

Habitat and distribution
Damp coniferous forests. Prefers poor, acid, marshy-peaty soil. Scattered and local, rarer in the South.

Additional information
The only plant in our flora where the flowers virtually always have 7 petals. With other species this is only coincidental and also very rare. An example of this is the Wood Anemone (p.85) but there can hardly be any confusion here.

White Butterbur
Petasites albus

Canadian Fleabane
Conyza canadensis

Daisy family *Asteraceae (Compositae)*
Flowering time: March–May .
Height of growth: 10–30 cm
Dicotyledonous; Perennial

Identification marks
Stout erect stem with light green leaf scales. Flowers in small narrow racemose capitula. Broad leaves appear only towards end of the flowering time; they are petiolate, round/heart-shaped, their underside is covered with white felt and the margins are irregularly spiny serrate.

Habitat and distribution
Woodlands, roadsides and waste places. Likes steep slopes and escarpments on saturated soils rich in nutrients. Local, from the Midlands and mid-Wales northwards.

Additional information
Without inflorescence similar: Butterbur (*P. hybridus*, p.243), leaves larger, grey-green on underside. Colt's-foot (*Tussilago farfara*, p.165): leaves smaller, with short, blackish teeth.

Daisy family *Asteraceae (Compositae)*
Flowering time: Aug.–Sept.
Height of growth: 8–100 cm
Dicotyledonous; Annual – Biennial

Identification marks
Flowers in numerous small capitula forming a panicled inflorescence. Stem erect, with bristly hairs, often branched, leaves alternate. Stem leaves linear-lanceolate, with bristly hairs.

Habitat and distribution
Weedy places in yards, railway track gravel, rubble heaps, paths, also on walls and fences. Likes gardens and clearings, too. Quite widespread but local, becoming rare in the north and Scotland. Likes nitrogen and warmth, hence does not do well in mountains.

Additional information
This plant was introduced around 200 years ago from North America and its former scientific name was *Erigeron canadensis*.

Heath Cudweed
Gnaphalium silvaticum

Gallant Soldier
Galinsoga parviflora

Daisy family *Asteraceae (Compositae)*
Flowering time: July–Sept.
Height of growth: 8–60 cm
Dicotyledonous; Perennial

Identification marks
Flowers in narrow-ovoid pointed capitula. Bracts grey-green with wide (usually) golden-brown shiny margins. Inflorescence a long, leafy spike. Stem erect, unbranched. Leaves alternate, lanceolate, hairy, white felt on the undersides, the uppersides glabrous.

Habitat and distribution
Light woodland, heaths, clearings, woodland ways, dry turf. Prefers soil which is low in lime, superficially acid and not too dry. Widespread and locally common.

Daisy family *Asteraceae (Compositae)*
Flowering time: May–Oct.
Height of growth: 10–75 cm
Dicotyledonous; Annual

Identification marks
Stem bushy-branching, more or less bare or with only a few short bristles. Leaves oval, pointed, saw-toothed; alternate. Globular capitula arranged in dichasial cymes; on the outside 4–6 white ligulate florets with 3-toothed apex, on the inside yellowish disc florets. Stems of capitula covered with short hairs with a few reddish-brown glandular hairs.

Habitat and distribution
Root-crop fields, gardens, and waste ground. Prefers porous nitrogenous soil in mild locations (vulnerable to frost). A fairly common weed in S. England.

Additional information
Similar: Shaggy Soldier (*G. ciliata = quadriradiata*), stem hairy, with spreading glandular and non-glandular hairs.

Scented Mayweed
Camomilla recutita

Scentless Mayweed
Matricaria perforata

Daisy family *Asteraceae (Compositae)*
Flowering time: June–July
Height of growth: 15–60 cm
Dicotyledonous; Annual

Identification marks
Flowers in paniculate capitula; on the outside white ligulate florets, on the inside yellow disc florets. Flower receptacle swollen, hollow, without scales. Stem erect, branched, glabrous. leaves bi- or tripinnate, ultimate segments long and narrow, less than 0.5 mm wide. Strong aromatic smell.

Habitat and distribution
Weedy places on fields and waysides. On loamy soils rich in nutrients. Local but often abundant in England and Wales. Also on sandy ground.

Additional information
Similar: Scentless Mayweed (adjacent) and Corn Chamomile (*Anthemis arvensis*), receptacle with scales among the disc florets. Broad leaves usually bipinnate, ultimate segments shorter, wider (0.5–1 mm). Arable fields. Locally common throughout.

Daisy family *Asteraceae (Compositae)*
Flowering time: uly–Sept.
Height of growth: 15–60 cm
Dicotyledonous; Annual

Identification marks
Flowers in paniculate capitula; on the outside white ligulate florets, on the inside yellow disc florets. Receptacle swollen, without scales. Stem erect, many branches at the top. Leaves divided into many fine lobes. Plant smells only slightly aromatic.

Habitat and distribution
Weedy places on rubbish tips, pathways and arable fields, also on railway ballast. Needs nitrogenous soils. Common throughout Britain.

Additional information
Also referred to as *Tripteurospermum inodorum* and *Matricaria inodora*. Similar (often combined to form one species): Sea Mayweed (*Matricaria maritimum*), leaves somewhat fleshy, stem rather prostrate. On salty ground.

Daisy
Bellis perennis

Stemless Carline Thistle
Carlina acaulis

Daisy family *Asteraceae (Compositae)*
Flowering time: Feb.–Nov.
Height of growth: 3–12 cm
Dicotyledonous; Perennial

Identification marks
All leaves in basal rosette, ovoid to spatulate, narrowing to a short wide petiole, margin usually notched. Capitula solitary on leafless scales, the ray florets are pure white or tinged with red on the back, the disc florets are yellow and cylindrical in shape.

Habitat and distribution
All types of short grassy areas: meadows, pastures, balks, parks, waysides and field-paths. Needs loamy soil, rich in nutrients and not too dry, in warm light locations. Very common.

Additional information
This old medicinal plant is also cultivated as an ornamental plant (f. *hortensis*) with many different varieties (usually filled, i.e. capitula contains only ligulate florets).

Daisy family *Asteraceae (Compositae)*
Flowering time: June–Sept.
Height of growth: 3–40 cm
Dicotyledonous; Perennial

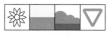

Identification marks
Leaves thorny, deeply pinnately spiny-lobed. Flowers in a single capitulum 4–7 cm in diameter. Only white to brownish-white cylindrical florets. Inner bracts linear, gleaming white, look like ray florets. Stem very short, prostrate to ascending, leafy.

Habitat and distribution
Not British. Semi-dry turf, heaths, pastures. Likes dry-warm calcareous stony ground. Scattered, in the north very rare, in the Alps up to over 2,000 m. Encouraged by grazing.

Additional information
Plants which have long stems or are heavily branched are only habitat-dependent varieties, like those without stems.

Yarrow
Achillea millefolium

Sneezewort
Achillea ptarmica

Daisy family *Asteraceae (Compositae)*
Flowering time: June–Aug.
Height of growth: 8–45 cm
Dicotyledonous; Perennial

Identification marks
Stem erect, leaves alternate, bi- or tripinnate, leaflets divided into 2–5 parts. Flowers in capitula, arranged in loose corymbs: on the inside yellowish-white disc florets, on the outside usually only 4–5 white or rarely red ray florets. Plant with aromatic smell.

Habitat and distribution
Common in meadows, semi-dry turf, balks, pastures, arable fields and along waysides and hedgerows. Prefers loose loamy soils which are rich in nutrients and not too moist.

Additional information
The colour of the ray florets varies from off-white through pure white to reddish pink and deep red.

Daisy family *Asteraceae (Compositae)*
Flowering time: July–Aug.
Height of growth: 20–60 cm
Dicotyledonous; Perennial

Identification marks
Flowers in capitula, arranged in loose corymbs; capitula up to 1.8 cm wide. On the outside wide, short ray florets, pure white to ivory-coloured, on the inside off-white disc florets. Stem erect, many leaves. Leaves undivided, lanceolate, serrate.

Habitat and distribution
Wet meadows, ditches, river banks. Prefers loamy soil which is occasionally saturated with ground water and not too poor in nutrients. Common, ascending to 800 m.

Additional information
Old medicinal plant.

Ox-eye Daisy
Leucanthemum vulgare

Scentless Feverfew
Tanacetum corymbosum

Daisy family *Asteraceae (Compositae)*
Flowering time: June–Aug.
Height of growth: 20–70 cm
Dicotyledonous; Perennial

Identification marks
Stem erect, little branching. At the end of each branch a single capitulum. The ray florets are long, narrow and white, the disc florets are yellow. Unpleasant smell. Lower leaves petiolate, crenate, upper ones sessile and serrate.

Habitat and distribution
Meadows, pathways, wasteland, balks; also found in light dry woodland and thickets. Very common, though less so in Scotland. On a variety of soils.

Additional information
Further names of this popular plant: Moon Daisy, Marguerite. Synonym: *Chrysanthemum leucanthemum*. Many forms.

Daisy family *Asteraceae (Compositae)*
Flowering time: June–Aug.
Height of growth: 50–100 cm
Dicotyledonous; Perennial

Identification marks
Usually 6–20 capitula in flat panicles. Capitulum 1–2 cm wide, outside narrow white ray florets, inside yellow disc florets. No scent. Stem stiff erect, few leaves especially at the top. Leaves tough, bipinnate. Pinnae often themselves coarsely toothed.

Habitat and distribution
Not British. Light mixed oak and beech woodland, thickets, forest margins, bushy heaths. Needs summer warmth. Likes calcareous soil rich in nutrients, not too moist. Scattered; very rare in the north.

Additional information
Difficult to classify systematically, therefore has a variety of names: *Chrysanthemum corymbosum*, *Leucanthemum corymbosum*, *Matricaria corymbosum*.

Marsh Helleborine
Epipactis palustris

Lesser Butterfly-orchid
Platanthera bifolia

Orchid family *Orchidaceae*
Flowering time: June–Aug.
Height of growth: 15–45 cm
Monocotyledonous; Perennial

Identification marks
Flower without spur, slightly drooping; lip clearly comprising 2 segments, edge waved and frilled, often has pink veins; remaining petals spread out, slightly greenish to brownish. Raceme unilateral. Stem erect. Leaves linear-lanceolate, sheathed, parallel venation.

Habitat and distribution
Fens, dune slacks; woodland in N. Europe. Local, sometimes frequent in England and Wales, as far north as C. Scotland. Needs ground which is at least occasionally wet, calcareous, contains humus.

Additional information
Similar: Other species, e.g. Broad-leaved Helleborine (*E. helleborine*, p.372), sometimes have a white lip.

Orchid family *Orchidaceae*
Flowering time: May–July
Height of growth: 15–45 cm
Monocotyledonous; Perennial

Identification marks
Stem erect, bearing 2 large oval broad leaves close to one another in the lower part and several small leaves above them. Many-flowered spike. Flowers night-scented. Spur is straight, slender and long, almost horizontal, only slightly curved at the end.

Habitat and distribution
Light deciduous or open forests, heaths and sunny balks. Scattered. Indicates slight surface acidity. Likes warmth, but can be found in (limestone) hills up to about 400 m.

Additional information
Very similar: Great Butterfly-orchid, *P. chlorantha*, on moister soils; in deciduous woods and on damp meadows. Spur thicker at the end, curved downwards. Flowers more greenish-white. Rather more common than its smaller relative.

White Helleborine
Cephalanthera damasonium

Creeping Lady's-tresses
Goodyera repens

Orchid family Orchidaceae
Flowering time: May–June
Height of growth: 15–80 cm
Monocotyledonous; Perennial

Identification marks
Flowers without spur, ivory in colour, ovoid as a result of the petals which press together, pointed. 3–12 flowers on spike. Leaves ovoid, spirally twisted, alternate.
Habitat and distribution
Woodlands. Prefers porous calcareous soils in none too cold location. Local in England only.
Additional information
Synonyms: C. grandiflora, C. alba, C. pallens, C. latifolia. Similar: Narrow-leaved Helleborine (C. longifolia = C. ensifolia), 3–15 pure white flowers, leaves narrow, 2-rowed. Woods. Local, rather rare, from S. England to Scotland.

Orchid family Orchidaceae
Flowering time: July–Aug.
Height of growth: 10–25 cm
Monocotyledonous; Perennial

Identification marks
Rhizome which creeps above ground. Leaves ovoid, thickish, with conspicuous network of veins; rosette formation at the base of erect flower stem. Further up the stem sheathing scale leaves, glandular-hairy like the flowers. Slightly spirally twisted, one-sided spike. Flowers small, whitish, covered with down, sweetish smell.
Habitat and distribution
Mossy coniferous forests. Prefers sandy soils, acid humus, not too damp and low in lime. Local becoming more rare, and disappearing from some parts: E. Anglia, N. England and Scotland.
Additional information
Also referred to as Satyrium repens.

Burnt Orchid
Orchis ustulata

Orchid family *Orchidaceae*
Flowering time: May–June
Height of growth: 8–20 cm
Monocotyledonous; Perennial

Identification marks
Dense globular to conical flower spike. Blossoms very small, about 5 mm long, initially brownish-red, later contrasted by the red-spotted white lip. The remaining petals are pressed together to form a helmet shape. Leaves are lanceolate.

Habitat and distribution
Widespread but local in England, on grassy hills and dry meadows. Prefers calcareous soil, poor in nutrients but warm and loamy.

Ghost Orchid
Epipogium aphyllum

Orchid family *Orchidaceae*
Flowering time: June–Aug.
Height of growth: 10–20 cm
Monocotyledonous; Perennial

Identification marks
Saprophytic plant with no chlorophyll. Stem erect, translucent reddish to yellowish; few leaf scales. Loose cluster of 1–4 flowers. These are drooping, whitish-yellowish-pale reddish, with spur and 3-lobed lip pointing upwards, wax-like and translucent.

Habitat and distribution
Oak and beech mixed woodland. Very rare. Prefers shady mossy soil which is well-moistened and contains nutrients and mull. Known only in a few localities in England and the Welsh borders.

Coral-root Orchid
Corallorhiza trifida

Corydalis cava

Orchid family *Orchidaceae*
Flowering time: May–Aug.
Height of growth: 7–25 cm
Monocotyledonous; Perennial

Identification marks
Pale green saprophytic plant. Stem narrow, erect, glabrous, usually with 3 somewhat inflated yellowish-green sheaths. No genuine leaves. Approximately 10 yellowish-white flowers without spurs, spreading and upright in lax clusters. Lip points downwards, 3-lobed, white with red spots or lines.

Habitat and distribution
Shady woodland, especially birch and pine, alder brake, peat bog spinneys. Likes decaying tree trunks. Rare, N. England and Scotland, in the mountains and occasionally on dunes. Calcifugous, avoids nitrogen. Decidedly a shady plant.

Additional information
In older works referred to under the name *C. innata*.

Fumitory family *Fumariaceae*
Flowering time: March–April
Height of growth: 10–20 cm
Dicotyledonous; Perennial

Identification marks
Stem erect, unbranched, with alternate broad leaves. These are bi- to tripinnate, blue-green, glabrous. Ultimate segments are ovoid, often repeatedly notched. 10–20 flowers form a dense raceme, bracts have entire margin. Flowers may be white to purple.

Habitat and distribution
A rare escape from gardens which may become established in a few places.

Additional information
Similar: Solid-tubered Fumitory (*C. solida*, see p.262).

Hare's-foot Clover
Trifolium arvense

White Clover
Trifolium repens

Pea family *Fabaceae (Leguminosae)*
Flowering time: June–Sept.
Height of growth: 10–40 cm
Dicotyledonous; Annual

Identification marks
Leaves trifoliate; leaflets narrow-oblong. Stem ascending to erect, branched. Flowers white, pink when fading; cylindrical inflorescence, sepals extending beyond petals, reddish and covered with feathery hairs.

Habitat and distribution
Dry turf, arable fields, open sandy ground, paths, wasteland. Only found on soils which are free of or low in lime, somewhat acid, loose and warm. Widespread throughout Britain but scattered and rather local.

Additional information
Weed and worthless as fodder, but as an old medicinal plant it had a variety of names.

Pea family *Fabaceae (Leguminosae)*
Flowering time: June–Sept.
Height of growth: 20–50 cm
Dicotyledonous; Perennial

Identification marks
Stem prostrate, rooted. Flowers in pedunculate, erect, round heads. Leaves stalked, pointing upwards, trifoliate; leaflets wedge-shaped to ovoid with almost heart-shaped apices, glabrous on the underside.

Habitat and distribution
Pastures, meadows, all types of turf – often sown. Very frequent. Resists trampling, likes nitrogen, reproduces well and therefore suitable for pastures, but also very hard to keep out of cultivated lawns. Up to over 900 m.

Additional information
Very many varieties; some are cultivated types which have escaped.

Mountain Clover
Trifolium montanum

White Melilot
Melilotus alba

Pea family *Fabaceae (Leguminosae)*
Flowering time: May–July
Height of growth: 15–40 cm
Dicotyledonous; Perennial

Pea family *Fabaceae (Leguminosae)*
Flowering time: July–Aug.
Height of growth: 60–120 cm
Dicotyledonous; Biennial

Identification marks
Stem erect-ascending, covered with woolly hairs, branched. Leaves trifoliate, toothed margin, underside hairy. Flowers in short globular heads, stalked, white or ivory-coloured.

Habitat and distribution
Not British. Balks, thickets, light dry woodland. Requires calcareous dry soil which is occasionally saturated. Must be low in nutrients. Scattered in limestone areas, in mountainous areas up to 1,800 m, absent in sandy areas.

Additional information
Distantly related: Alsike Clover (*T. hybridum*), glabrous or only slightly hairy, leaflets ovoid, flowers first white, later pink. Scattered; damp meadows, pathways, wasteland.

Identification marks
Stem erect, many branches. Many long narrow slightly unilateral racemes, erect and bearing large number of flowers. Leaves trifoliate; leaflets ovoid, serrate. At base of leaf-stalk 2 bristle-like stipules.

Habitat and distribution
Weedy places along waysides, slopes, railway embankments, gravel banks. Needs rather dry ground rich in nutrients and often stony, but also found on pure loamy soil in warm locations. Frequent in S. England and Wales, but less so elsewhere.

Additional information
Poor fodder but valuable plant for bees.

Wood Vetch
Vicia sylvatica

White Deadnettle
Lamium album

Pea family *Fabaceae (Leguminosae)*
Flowering time: June–Aug.
Height of growth: 60–130 cm
Dicotyledonous; Perennial

Identification marks
Leaves pinnate, with 5–10 pairs of ovoid leaflets and branched, terminal climbing tendrils. White flowers with mauve veins, around 1.5 cm long, drooping. Many-flowered erect raceme. Stem quadrangular in cross-section.

Habitat and distribution
Light deciduous and mixed woodland, wood margins, thickets, grassy areas. Found on dry, warm, loose loamy soils rich in lime and nutrients and often stony. Common throughout Britain.

Additional information
Similar: Wood Bitter Vetch (*V. orobus*); without tendrils, stem erect, rarely taller than 0.5 m, hairy. Local in woods.

Mint family *Lamiaceae (Labiatae)*
Flowering time: May–Dec.
Height of growth: 10–50 cm
Dicotyledonous; Perennial

Identification marks
Looks like a nettle, but without the stinging hairs. 5–8 bilabiate flowers in axillary whorls (verticillasters). Stem quadrangular, erect. Leaves opposite and decussate, ovoid, dentate; petiolate.

Habitat and distribution
Waysides, waste ground, railway embankments, walls, fences, thickets. Likes nitrogenous soil, thus frequently found on fertilized ground but rarely in fields. Very common in England and Wales, rare in N. Scotland.

Additional information
May be confused with related plants with yellow and red flowers (pp.192 and 274f. respectively). White Deadnettle always has 3–7 cm long leaves with long pointed teeth and without flecks on them.

Yellow Woundwort
Stachys recta

Annual Woundwort
Stachys annua

Mint family *Lamiaceae (Labiatae)*
Flowering time: June–Oct.
Height of growth: 10–30 cm
Dicotyledonous; Perennial

Identification marks
Stem erect or ascending, quadrangular. Leaves opposite and decussate, with very short stalks, narrow ovoid. Inflorescence made up of individual whorls where the flowers get smaller towards the top. Flowers yellowish-white, approx. 1.5 cm long; 6–10 per whorl.

Habitat and distribution
Naturalized in one locality – in Barry, S. Wales. Rare in N. Europe. Likes stony calcareous ground. Not found in silicate areas or in the mountains over 1,000 m.

Additional information
Old medicinal and magic plant with a variety of popular names. Similar: *S. annua* (see adjacent photo).

Mint family *Lamiaceae (Labiatae)*
Flowering time: June–Oct.
Height of growth: 10–30 cm
Dicotyledonous; Annual

Identification marks
Stem usually erect and branched, quadrangular. Leaves opposite and decussate, narrow ovoid, short-stalked, the upper one sessile. Inflorescence made up of individual flower whorls. Flowers yellowish-white, around 1–1.5 cm long; 2–6 per whorl.

Habitat and distribution
Arable fields and waste places. Likes calcareous and nitrogenous loamy soils in rather warm dry locations. Occurs as a casual in many places but does not become established.

Bastard Balm
Melittis melissophyllum

Hedge Hyssop
Gratiola officinalis

Mint family *Lamiaceae (Labiatae)*
Flowering time: May–July
Height of growth: 20–50 cm
Dicotyledonous; Perennial

Identification marks
Stem quadrangular, erect, little branching. Broad leaves opposite and decussate, stalked, ovoid, conspicuously wrinkled, coarsely round-toothed margin. Entire plant has dense covering of soft hairs. Flowers few in number in axils of upper leaves, often all favouring one side; smells of honey.

Habitat and distribution
Light deciduous forests, woods and hedgerows. Prefers loose calcareous soils, warm but not too dry. Local to rare in Wales and southern parts of England.

Additional information
The flowers are usually white spotted with pink but may sometimes be more or less completely pink.

Figwort family *Scrophulariaceae*
Flowering time: June–Aug.
Height of growth: 10–50 cm
Dicotyledonous; Perennial

Identification marks
Stem erect to ascending, round, hollow. Leaves opposite and decussate, sessile, lanceolate, margin serrate. Entire plant glabrous. Flowers solitary in leaf axils, pedicellate. Corolla often has reddish veins or the upper lip has a reddish tinge.

Habitat and distribution
Not British. In reeds in stagnant or slow-flowing waters, on river banks, ditches and damp meadows. Prefers muddy calcareous and dense soils. Withstands summer drought and excessive salt. Only found on lower levels. Very rare.

Additional information
Old (poisonous) medicinal plant.

Eyebright
Euphrasia rostkoviana

Alpine Butterwort
Pinguicula alpina

Figwort family *Scrophulariaceae*
Flowering time: July–Aug
Height of growth: 10–40 cm
Dicotyledonous; Annual

Identification marks
Stem ascending, usually little branched. Leaves opposite and decussate, ovoid, small, coarsely round-toothed. Inflorescence has leaflike bracts. Leaves, bracts and calyx are glandular-hairy. Flowers usually about 1 cm long, white with yellow, black-mauve, bluish or reddish spots on the throat of the corolla.

Habitat and distribution
Meadows, pastures; tends to avoid lime and reacts badly to fertilizer. Local to rare in N. England, Wales, and Border region.

Additional information
Many different species, hard to determine the limits of the various species in the genus.

Butterwort family *Lentibulariaceae*
Flowering time: May–July
Height of growth: 5–15 cm
Dicotyledonous; Perennial

Identification marks
Leaves in basal rosette, yellowish-green, lanceolate with upturned margins, sticky on the upperside. Flowers single, stalked, drooping horizontally, spur is curved downwards. Throat of the corolla is wide open and has yellowish flecks on it.

Habitat and distribution
Probably extinct in Britain. Flat moorland, areas round springs, Alpine mats which are irrigated or saturated by ground water, rocky fissures. Prefers wet calcareous ground. Scattered up to 2,300 m in Alpine regions, very rare in the area of the Lower Alps.

Additional information
Blue flowering Common Butterwort, *P. vulgaris* (p.347), occasionally produces blossoms with white spots. Even when both species adjacent there is no hybridizing.

Traveller's Joy
Clematis vitalba

Holly
Ilex aquifolium

Buttercup family *Ranunculaceae*
Flowering time: July–Aug.
Height of growth: up to 30 m
Dicotyledonous

Identification marks
Liana with woody climbing stem. Leaves opposite, pinnate; leaflets ovoid/heart-shaped; petioles tendril-like. Terminal and axillary panicles. Petals absent, sepals petal-like. Fruit has long feathery-hairy style.

Habitat and distribution
Hedges, wood margins, thickets. Prefers soil rich in chalk and nutrients. Likes nitrogenous ground, is thus often found near villages. England and Wales as far north as the Midlands and S. Yorkshire. Likes the warmth so is only found in moderately hilly areas.

Holly family *Aquifoliaceae*
Flowering time: May–Aug.
Height of growth: 3–15 m
Dicotyledonous

Identification marks
Evergreen, shiny leaves, ovoid, 3–10 cm long, toothed with prickles, slightly waved. Petiole very short. Flowers small, in axillary clusters. Berries round, shiny and red.

Habitat and distribution
Deciduous and coniferous woods. On all but very wet soils. Very shade-tolerant. Sometimes found in the garden as a hedge or as an ornamental.

Additional information
Similar in its foliage: Oregon Grape (*Mahonia aquifolium*), yellow flowers, blue fruit. Decorative shrub, rarely naturalized.

Dogwood
Cornus sanguinea

Wild Privet
Ligustrum vulgare

Dogwood family *Cornaceae*
Flowering time: May–July
Height of growth: 0.25–4 m
Dicotyledonous

Identification marks
Twigs often have a reddish tinge (blood-red in autumn and winter). Flowers in flat corymbose cymes. Leaves opposite, with curved pinnately branched venation, ovoid, margin entire. Fruit is berry-like, globular, black.

Habitat and distribution
Deciduous and mixed woodland, hedges. Prefers calcareous soil rich in nutrients and not too damp in sunny, warm location. Common in England and Wales, rarer in Scotland, where it is introduced.

Additional information
Has been listed under the genus *Swida* (*S. sanguinea*). Varieties with mottled leaves cultivated (rare; other dogwood species are preferred).

Olive family *Oleaceae*
Flowering time: June–July
Height of growth: under 5 m
Dicotyledonous

Identification marks
Leaves opposite and decussate, leathery, ovoid-lanceolate, margin entire and glabrous. Short petiole. Many-flowered panicles. Flowers funnel-shaped, with an unpleasant smell. Berries small, globular, shiny and black, very juicy.

Habitat and distribution
Thickets, hedges, wood margins. Likes loose calcareous soils, often somewhat stony. Common in England and Wales, introduced in Scotland. Often found in gardens as a hedge, can be cut. Only relative resistance to frost.

Additional information
In warmer regions the leaves remain green all winter. In gardens it is often replaced by Garden Privet (*L. ovalifolium*), susceptible to frost.

Pear
Pyrus communis

Crab Apple
Malus sylvestris

Rose family *Rosaceae*
Flowering time: April–May
Height of growth: 5–15 m
Dicotyledonous

Identification marks
Twigs are glabrous or rapidly become so. Spiny at the tips or on the short lateral twigs. Leaves petiolate, round-ovoid, margin is finely serrated, shiny on the upperside, tough-leathery. Umbel-like corymbs with few flowers at the end of the short shoots. Anthers purple red. Fruit is a small woody pear. Often grows as a large shrub.

Habitat and distribution
Typically in hedgerows, usually found as individual trees rather than in large numbers. Widespread in England and Wales but not in large numbers. Prefers loamy soils rich in lime and nutrients. Often cultivated from wild.

Additional information
Pirus is an old spelling for the generic name.

Rose family *Rosaceae*
Flowering time: May
Height of growth: 2–10 m
Dicotyledonous

Identification marks
Twigs sometimes thorny. Leaves petiolate, broadly ovoid, often with off-centre apex, underside glabrous. Corymbs with few flowers. Anthers yellow, petals white or pink, 1–3 cm long. Fruit is a small apple 2–3 cm in diameter, dry and sour, somewhat woody.

Habitat and distribution
Deciduous woodland, light thickets, hedgerows. Likes calcareous well-moistened soils rich in nutrients. Common in England and Wales, becoming rarer in Scotland.

Additional information
Similar to the Crab Apple and partially a hybrid of this is the cultivated apple (*M. domestica*) with its many different varieties. Old types grown wild are hard to distinguish. They generally have leaves with felt-like undersides.

Medlar
Mespilus germanica

Snowy Mespil
Amelanchier ovalis

Rose family *Rosaceae*
Flowering time: May–June
Height of growth: 2–3(6) m
Dicotyledonous

Identification marks
Thorny twigs, covered with felt-like hairs when young. Leaves lanceolate, finely serrated, undersides covered with soft hairs. Flowers solitary and terminal on twig tips; the narrow herbaceous calyx lobes are longer than the petals and later crown the rough, leather-brown fruit. Usually grows as a shrub.

Habitat and distribution
Hedgerows. Rare. Naturalized in S. England, Midlands and as far north as S. Yorkshire. On dry, warm calcareous ground.

Additional information
In the Middle Ages, cultivated for its edible fruit.

Rose family *Rosaceae*
Flowering time: April–May
Height of growth: 1–3 m
Dicotyledonous

Identification marks
Small shrub, twigs without thorns. Leaves elliptical, serrate, 2–4 cm long, petiolate. Racemes few-flowered. Petals narrow, 1–2 cm long, covered with tufts of hair on the outside. Fruit the size of a pea, globular, blue-black.

Habitat and distribution
Rocky slopes, stony forest margins, sunny thickets. Very rare. Found on stony, flat, dry ground, preferably calcareous. In sunny location up to well over 1,500 m.

Additional information
A North American shrub, planted in parks and gardens, not yet escaped or naturalized in Britain.

Wild Service-tree
Sorbus torminalis

Common Whitebeam
Sorbus aria

Rose family *Rosaceae*
Flowering time: May–June
Height of growth: under 5 m
Dicotyledonous

Identification marks
Medium-sized tree. Leaves 7–11 lobed, upperside glabrous, underside grey and felt-like when young, later glabrous. Lobes are pointed, coarsely serrate; the lower ones wide-spreading or at right angles to the petiole, the upper ones less so. Flowers in compound corymbs. Petals approx. 5 mm long. Fruit obovoid, brown, covered with light-coloured spots. Autumn foliage conspicuously red in colour.

Habitat and distribution
In sunny deciduous and mixed woodland which is warm in summer and relatively dry. Scattered through England and Wales. Calcicolous, likes nitrogenous soil.

Rose family *Rosaceae*
Flowering time: May–June
Height of growth: under 15 m
Dicotyledonous

Identification marks
Shrub or small tree. Undersides of leaves are covered with white felt, doubly serrate or slightly lobed. Umbrella-like corymbs. Petals approx. 5 mm long. Fruit ovoid-globular; orange-yellow to red in colour.

Habitat and distribution
Dry woodland, stony thicket-covered slopes. Usually on calcareous stony ground. Native in S. England but often planted elsewhere.

Additional information
Many varieties and often hybrids with other species. A number of closely related and difficult to distinguish species of *Sorbus* are considered to be endemic to the British Isles.

Rowan
Sorbus aucuparia

Hawthorn
Crataegus monogyna

Rose family *Rosaceae*
Flowering time: May–June
Height of growth: 5–15 m
Dicotyledonous

Identification marks
Usually a small tree. Leaves unpaired pinnate; 9–17 oblong leaflets on short stalks with finely serrate margins. Many-flowered corymbs. Petals approx. 5 mm long. Fruit globular, the size of a pea, red.

Habitat and distribution
Woodland, thickets, heaths, mountains. Found on dry and moist slightly acid soil, usually free of lime and low in nutrients. Common in north and west, becoming rare in eastern England and parts of the Midlands. Up to about 1,000 m.

Additional information
Is also widely referred to as Mountain Ash. Very many hybrids and varieties.

Rose family *Rosaceae*
Flowering time: May–June
Height of growth: 2–10 m
Dicotyledonous

Identification marks
Usually a shrub or small tree with light grey bark and thorny twigs. Leaves deeply lobed; usually 3–5 lobes, pointed, serrate at the front. Erect lax corymbs. Flowers have strong smell. One style per flower, 1 stone in each red oval fruit. Pedicel hairy.

Habitat and distribution
Woods, thickets, hedges. In shallow soil which can be stony. Withstands drought. Very common in almost all parts except N. Scotland. Often used as hedging.

Additional information
Similar: Midland Hawthorn (*C. laevigata* = *C. oxyacantha*), pedicel usually glabrous, flower has 2–3 styles, fruit 2–3 stones. Leaves slightly lobed. Somewhat local, mainly in eastern England.

Field Rose
Rosa arvensis

Wild Cherry
Prunus avium

Rose family *Rosaceae*
Flowering time: June–July.
Height of growth: 50–100 cm
Dicotyledonous

Identification marks
Stems green, sometimes tinged purple, curved prostrate or creeping, rarely climbing over another shrub; many thorns. Leaves pinnate, 5–7 leaflets; leaflets ovoid, toothed. Flowers solitary, on pedicels, 3–5 cm wide. Styles united to form a conspicuous column. Small hip.

Habitat and distribution
Light- and mixed-deciduous woodland, woodland margins, woodland paths, glades, also rocky hedges. Found on slightly acid ground, not too dry but rich in humus. Common in the south, rarer towards the north, very rare in Scotland.

Additional information
The creeping growth and the protruding styles distinguish it clearly from other wild roses.

Rose family *Rosaceae*
Flowering time: Feb.–June
Height of growth: 5–25 m
Dicotyledonous

Identification marks
Tree, twigs greyish-brown. Leaves usually appear together with the flowers, obovoid, pointed apex, toothed margin. Petiole has 1–2 brown half-spherical glands at the top. Flowers in umbels of 2–6, long pedicels, 2–3 cm wide. Fruit: small cherry.

Habitat and distribution
Woods, hedges, copses, thickets. On well-moistened soils rich in nutrients. Common in England and Wales, becoming rare in Scotland.

Additional information
Under the name *Cerasus avium* is attributed to a different genus. The Garden Cherry is similar; may have been derived from Wild Cherry by hybridization.

Bird Cherry
Prunus padus

Rose family *Rosaceae*
Flowering time: May
Height of growth: 3–15 m
Dicotyledonous

Identification marks
Shrub or small tree. Flowers in many-flowered pendulous racemes at the end of leaf twigs; strong smell. Leaves alternate, ovoid, pointed at the apex, somewhat wrinkled, margin doubly serrate. Fruit the size of a pea, cherry-like, black.

Habitat and distribution
Moist deciduous woodland, thickets close to the water. Found on deep loamy soil with abundant nutrients. Common in Scotland, N. England and Wales, as far south as the Midlands. Sometimes planted in S. England.

Blackthorn
Prunus spinosa

Rose family *Rosaceae*
Flowering time: March–May
Height of growth: 1–4 m
Dicotyledonous

Identification marks
Gnarled shrub with blackish bark and thorny twigs. Flowers usually appear before the leaves, solitary (but numerous) on short pedicels. Leaves elliptical, margin sharply serrate. Fruit globular, about 1 cm in length, with bluish bloom, black, dry-sour.

Habitat and distribution
Thickets, woodland margins, hedgerows, path verges. Found on ground which is not too dry but rich in minerals. Throughout most of Britain up to about 450 m.

Additional information
Together with the related Plum the Sloe sometimes forms the hybrid *P.* x *fruticans*, which baffles anyone trying to classify it. It looks like a Sloe, has no thorns and the fruit is large and sweet.

Raspberry
Rubus idaeus

Bramble
Rubus fruticosus

Rose family *Rosaceae*
Flowering time: June–Aug.
Height of growth: 100–160 cm
Dicotyledonous

Identification marks
Stem covered with fine prickles, erect or arching. Leaves alternate, undersides have thick covering of white hairs, uppersides light green, 3–7 leaflets which are ovoid, upperside wrinkled, toothed margin. Somewhat drooping cyme with few flowers. Compound fruit red.

Habitat and distribution
Glades, clearings, woodland margins, heaths, especially in hilly areas. Likes nitrogen. Common. Up to 1,000 m. Sometimes occurs in large numbers.

Additional information
Not always easy to distinguish from the other species of *Rubus* (see p.61 and adjacent entry) when it is in flower. All characteristics should be examined.

Rose family *Rosaceae*
Flowering time: May–Sept.
Height of growth: 20–200 cm
Dicotyledonous

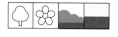

Identification marks
Stem usually covered with coarse prickles, arching, ascending or creeping. Leaves alternate, ternate or palmately divided into 3–(5)7 leaflets, undersides glabrous or covered with white felt. Inflorescence rather variable. (Compound) fruit black (red).

Habitat and distribution
Woodland, hedgerows, wayside verges, fields, gardens, fallow land, heaths. Very common throughout Britain. Usually found on soils which are not too dry but low in lime and containing nutrients.

Additional information
Sometimes divided into dozens of hard-to-distinguish variants. The fruit and stems of the Dewberry (*R. caesius*) have a bluish tinge. Likes damp habitats; common in England and Wales, less so in Scotland.

Bog Bilberry
Vaccinium uliginosum

Cowberry
Vaccinium vitis-idaea

Heath family *Ericaceae*
Flowering time: May–June
Height of growth: up to 50 cm
Dicotyledonous

Identification marks
Flat prostrate or erect to ascending shrub with many branches. Bark greyish-brown. Leaves blue-green, obovoid, blunt, margin entire, very short petioles. Several flowers in each leaf axil, drooping, campanulate. Fruit is a globular berry with a blue bloom and colourless juice.

Habitat and distribution
Moorland, marshy woodland and heaths. On damp to wet acid soil. Local but sometimes abundant on Bilberry moors.

Heath family *Ericaceae*
Flowering time: June–Aug.
Height of growth: 10–30 cm
Dicotyledonous

Identification marks
Leaves leathery, evergreen; rolled up at the edge. Several flowers in terminal racemes, pink or pure white, slightly drooping. Flowers campanulate, usually 5, rarely 4 fused petals. Fruit: a berry, first white, then shining red when ripe, in dense clusters, usually unilateral.

Habitat and distribution
In mixed and coniferous woodland, high moorland, heaths with stunted bushy growth. Common in the mountains and sometimes becoming dominant. Requires acid, meagre soil saturated at intervals and containing coarse humus.

Additional information
Similar: Bearberry (*Arctostaphylos uva-ursi*), especially when this is not in flower. The leaf edges are flat (see p.114).

Ledum palustre

Bearberry
Arctostaphylos uva-ursi

Heath family *Ericaceae*
Flowering time: June–July
Height of growth: up to I m
Dicotyledonous

Identification marks
Unpleasant smelling shrub, twigs covered with red felt. Leaves narrow lanceolate, up to about 0.5 cm wide, leathery, evergreen, undersides covered with rust-red hairs. Many-flowered terminal umbel-like racemes. Flowers spread out like rays, corolla often over 1 cm in diameter.

Habitat and distribution
Boggy areas. Very rare in a few areas of lowland Scotland. Needs wet peaty soil free of lime and low in nutrients.

Additional information
Labrador-tea (*L. groenlandicum*), from North America, is a rare escape from cultivation. Leaves wide lanceolate to ovoid, upperside bumpy and rough.

Heath family *Ericaceae*
Flowering time: May–July
Height of growth: 20–60 cm
Dicotyledonous

Identification marks
Prostrate shrub with ascending twigs. Leaves obovate, glabrous, leathery, evergreen, with smooth flat edges. 5–12 globular-campanulate flowers in terminal racemes; sometimes tinged with pink. Fruit red, floury.

Habitat and distribution
High moors. Needs soil which is rich in humus and has at least surface acidity but which must be warm and rather dry. Common in N. England and Scotland, also on high ground in the Midlands.

Elder
Sambucus nigra

Wayfaring-tree
Viburnum lantana

Honeysuckle family *Caprifoliaceae*
Flowering time: June–July
Height of growth: 3–10 m
Dicotyledonous

Identification marks
Shrub with alternate leaves, often long off-shoots from the base; bark has protruding pores; pith is white. Flat umbel-like flowers with 5 main rays, erect, begins to droop when ripe. Rays also turn red. Black berries. Leaves pinnate; 3–7 large serrate leaflets.

Habitat and distribution
Woods, glades, hedges, thickets. On disturbed soil rich in humus. Likes nitrogen. Moisture indicator. Common; often found on neglected areas; less so in Scotland.

Additional information
Old medicinal and berry fruit plant. Sometimes found as an ornamental bush with bipinnate slit foliage (var. *laciniata*).

Honeysuckle family *Caprifoliaceae*
Flowering time: May–June
Height of growth: 2–6 m
Dicotyledonous

Identification marks
Shrub with many branches; when young the twigs are covered with grey felt-like hairs. Leaves opposite, elliptical, finely serrate margin, undersides wrinkled, with grey felt. Petioles short. Flat terminal umbel-like cymes. Fruit berry-like and red, black in the final stage, ovoid, pressed together laterally.

Habitat and distribution
Deciduous and mixed woodland, thickets. Requires sunny, warm loose calcareous soil. Common in S. England, rarer northwards and westwards into Wales. Introduced in Scotland. Scattered; absent in sandy regions.

Guelder-rose
Viburnum opulus

Fly Honeysuckle
Lonicera xylosteum

Honeysuckle family *Caprifoliaceae*
Flowering time: June–July
Height of growth: 2–4 m
Dicotyledonous

Identification marks
Shrub with opposite leaves. Glabrous twigs. Leaves 3–5 lobes. Base of petiole has stipules resembling bristles and cup-like glands. Flat terminal umbel-like cymes. Outer flowers sterile, enlarged. Fruit berrylike, red.

Habitat and distribution
Moist deciduous and mixed woodland, scrub and riverbank thickets. On loamy soils saturated by ground water. Moisture indicator. Common in England and Wales, less so in Scotland.

Additional information
The Snowball Tree (var. *roseum*) is often grown as an ornamental shrub. It has globular heads composed of enlarged sterile flowers. Similar in its leaves: Maple species, e.g. Sycamore (*Acer pseudoplatanus*, p.381), glands and stipules are absent on petiole.

Honeysuckle family *Caprifoliaceae*
Flowering time: May–June
Height of growth: 1–2 m
Dicotyledonous

Identification marks
Branched bushy shrub with hollow, rod-shaped twigs. Flowers always in pairs on the peduncle, the ovaries are fused at the base. Corolla hairy. Leaves opposite, margin entire, broadly ovoid. Shiny red paired berries.

Habitat and distribution
Deciduous woodland, in hedges. Calcicolous. Needs loose soil rich in humus and nutrients. Scattered in England and Wales, and a few places in Scotland. Introduced in most of these localities.

Additional information
The flowers are never completely white. In parks and gardens the very similar *L. ruprechtiana* (originates in China) is often planted. Its flowers are snow-white and have no hair on the outside.

Honeysuckle
Lonicera periclymenum

False Acacia
Robinia pseudoacacia

Honeysuckle family *Caprifoliaceae*
Flowering time: June–Sept.
Height of growth: up to 6 m
Dicotyledonous

Identification marks
Woody climber or twining shrub. Leaves opposite, ovoid, the upper ones sessile, the lower ones have short petioles. Terminal whorls of flowers. Flowers scented, cloudy white, often tinged with pink.

Habitat and distribution
Deciduous and mixed woodland, forest margins, thickets. Somewhat calcifugous. Likes mild winter climate. Common.

Additional information
Similar: Perfoliate Honeysuckle, *L. caprifolium*, upper leaves fused in pairs to form oval or round discs with the stem passing through the centres. Old ornamental plant from the Eastern Mediterranean region, often found growing wild in hedgerows in S. and E. England, also in a few parts of the north.

Pea family *Fabaceae (Leguminosae)*
Flowering time: June
Height of growth: 10–27 m
Dicotyledonous

Identification marks
Tree with light brown bark that has deep longitudinal grooves. Shiny, reddish-brown twigs covered with large double thorns. Leaves pinnate; 9–19 ovoid leaflets, somewhat greyish-green in colour. Many-flowered pendulous racemes. Flowers are scented. Reddish-brown pod.

Habitat and distribution
Popular tree in parks and council planting schemes, rarely escaping. Likes warmth.

Additional information
The tree was first introduced by J. Robin in 1601 and was believed to be an acacia, hence the current name 'False Acacia'.

Branched Bur-reed
Sparganium erectum

Common Meadow-rue
Thalictrum flavum

Bur-reed family *Sparganiaceae*
Flowering time: June–Aug.
Height of growth: 50–150 cm
Monocotyledonous; Perennial

Identification marks
Leaves grass-like, somewhat stiff. Stem branched. Male and female flowers are in separate round heads at the ends of the branches, the male ones at the top and the female ones lower down the stem; fruit is prickly, resembling a bur (hence the name).

Habitat and distribution
Frequently found in reed-beds in stagnant or slow-flowing water, in ditches and marshes. Likes soil containing nutrients. Common in almost all areas.

Additional information
Similar rarer species where the stem is not branched: Unbranched Bur-reed (*S. emersum*), Floating Bur-reed (*S. angustifolium*) with floating leaves, Least Bur-reed (*S. minimum*) with 2–5 small flower heads.

Buttercup family *Ranunculaceae*
Flowering time: July–Aug.
Height of growth: 50–100 cm
Dicotyledonous; Perennial

Identification marks
Stem erect, glabrous, usually unbranched. Leaves alternate, the lower ones petiolate, the upper ones sessile, bi- or tripinnate. Flowers erect, in bushy clustered panicles, sweet-smelling; petals wither rapidly; many stamens, yellow.

Habitat and distribution
Fens, damp meadows, river banks. Likes wet loamy soil which is dry in summer. Frequent, as far north as Inverness.

Greater Celandine
Chelidonium majus

Drooping Bitter-cress
Cardamine enneaphyllos

Poppy family *Papaveraceae*
Flowering time: May–Aug.
Height of growth: 30–90 cm
Dicotyledonous; Perennial

Identification marks
The plant contains an orange-yellow milky juice (latex). Flowers are in clusters or solitary in leaf axils. Leaves almost pinnate, coarsely lobed or toothed, undersides bluish-green. Plant glabrous or with scattered hairs.

Habitat and distribution
On rubble areas, pathways, along walls and fences, in gardens, damp light woodland and thickets, forest margins. Nitrogen indicator. Likes warmth but not full sunshine. Common, especially near habitation.

Mustard family *Brassicaceae (Cruciferae)*
Flowering time: April–May
Height of growth: 20–30 cm
Dicotyledonous; Perennial

Identification marks
Stem erect but at an angle, no leaves lower down. At the top is a whorl of 3 triple-fingered leaves with short petioles. Flowers in terminal clusters but drooping and thus hanging below the leaf whorl. Flowers pale yellow, 1–2 cm long. Fruits are narrow, 5–8 cm long but erect. When fruit is borne the plant often has long-stemmed basal leaves.

Habitat and distribution
Not British. Deciduous and mixed deciduous woodland. On well-moistened loamy soil, rich in nutrients and mull. Rare. Main area of distribution is south-east Germany: the Alps and extensive foreland.

Additional information
Other scientific name: *Dentaria enneaphyllos*.

Charlock
Sinapis arvensis

Mustard family *Brassicaceae (Cruciferae)*
Flowering time: May–July
Height of growth: 30–80 cm
Dicotyledonous; Annual

Identification marks
Racemes initially short, later becoming extremely long. Flowers deep yellow, 1.5–2 cm wide; narrow, horizontally projecting sepals, yellowish-green in colour. Stem usually erect and unbranched. Leaves undivided but often pronouncedly crenate, the lower ones almost lyre-shaped. Fruits much longer than they are wide.

Habitat and distribution
Fields, gardens, wasteland. On calcareous soil rich in nutrients. Common.

Additional information
Very similar: Wild Radish, *Raphanus raphanistrum* (p.38), type which has light yellow flowers; often found in the same habitats but the sepals are erect.

Winter-cress
Barbarea vulgaris

Mustard family *Brassicaceae (Cruciferae)*
Flowering time: May–Aug.
Height of growth: 30–90 cm
Dicotyledonous; Perennial

Identification marks
Basal leaves in rosettes, lyre-shaped with small round terminal lobes; stem leaves pinnate, upper ones undivided, sessile, clasping the stem and projecting. Stem angular, usually branched. Inflorescence dense to begin with, elongating later. Petals twice as long as sepals. Ovary long, quadrangular.

Habitat and distribution
Weedy areas along pathways, railway embankments, river banks, on gravel banks and sandbanks, waste areas, in clearings. Likes well-moistened stony ground rich in nutrients. Common but less so in north.

Additional information
Similar: Small-flowered Winter-cress (*B. stricta*), sepals c. 1/3 length of petals; Medium-flowered Winter-cress (*B. intermedia*), leaves pinnate.

Hedge Mustard
Sisymbrium officinale

Treacle Mustard
Erysimum cheiranthoides

Mustard family *Brassicaceae (Cruciferae)*
Flowering time: June–July
Height of growth: 30–90 cm
Dicotyledonous; Annual

Identification marks
Inflorescence initially corymb-like but elongating later. Petals 2–3 mm long. Fruit approx. 1 cm long, narrow, pointed, pressed closely to the stem. Leaves pinnate with rounded terminal lobe, upper leaves with a spear-shaped terminal lobe.

Habitat and distribution
A weed of arable land. Pathways, rubble, walls, railway embankments, river banks. On warm soil which is not too damp. Nitrogen indicator. Frequent.

Additional information
The following also have fruit pressed close to the stem: Tower Mustard (*Arabis glabra*), flowers light yellow, leaves glabrous, bluish-green. Black Mustard (*Brassica nigra*), petals about 1 cm long. *Eruca sativa*, petals about 1.5–2 cm long.

Mustard family *Brassicaceae (Cruciferae)*
Flowering time: June–Aug.
Height of growth: 15–90 cm
Dicotyledonous; Annual

Identification marks
Flowers small, on relatively long pedicels. Sepals erect, flowers barely 6 mm wide. Inflorescence corymb-like or racemes. All leaves undivided, very coarsely serrate, lanceolate, narrower at the base. Fruit at least 10 times longer than wide.

Habitat and distribution
Fields, gardens, pathways, riverbanks, gravel banks and sand banks. Grows on loose slightly damp soil which is somewhat calcareous. Local, sometimes common at low altitudes in the south, rarer in the north.

Additional information
Yellow cruciform flowers, lanceolate leaves, erect calyx and fruits indicate that this belongs to genus *Erysimum*, if petals less than 2 cm long. Wallflower (*Cheiranthus cheiri*) very similar but flowers over 2 cm wide.

Small Alison
Alyssum alyssoides

Alyssum montanum

Mint family *Brassicaceae (Cruciferae)*
Flowering time: May–June
Height of growth: 5–25 cm
Dicotyledonous; Annual

Identification marks
Bushy plant with erect or curved ascending stems. Leaves obovoid to narrow-oblanceolate, margin entire, with grey felt-like down. Flowers in dense racemes, approx. 3 mm wide, pale yellow fading to white, with persistent sepals. Small siliculae round, flattened, covered with rough hairs.
Habitat and distribution
Fields, both arable and pasture. In sunny dry habitats. Likes lime and a small amount of nitrogen. Scattered in southern and eastern England and parts of eastern Scotland.
Additional information
Synonym: *A. calycinum*.

Mustard family *Brassicaceae (Cruciferae)*
Flowering time: April–June
Height of growth: 5–25 cm
Dicotyledonous; Perennial

Identification marks
Lower part of stem quite woody, many branches. Leaves have entire margin, with grey felt-like covering, narrow, up to 2 cm long. Dense racemes. Flowers golden-yellow, about 5 mm wide. Sepals wither rapidly. Small siliculae oval to round, flattened, grey felt-like covering.
Habitat and distribution
Not British. Rocks and dry turf. Needs sunny, warm, sandy or stony ground, dry and calcareous. Very rare but it forms small groups in those localities where it does occur.
Additional information
Similar: Golden Alison (*A. saxatile*), leaves up to 5 cm long; many-flowered panicle; small siliculae glabrous. Not British. Very rare in rocky areas but widely spread in rockeries as ornamental plant.

Woad
Isatis tinctoria

Ball Mustard
Neslia paniculata

Mustard family *Brassicaceae (Cruciferae)*
Flowering time: July–Aug.
Height of growth: 50–120 cm
Dicotyledonous; Biennial – Perennial

Identification marks
Very densely branched, at least at the top, usually with many flowers. Leaves on the stem bluish-green, glabrous, margin usually entire, heart-shaped or sagittate clasping the stem. Fruit flat, broader towards the apex, broadly winged; pendulous, ultimately black-brown-violet, usually 1-seeded.

Habitat and distribution
Cultivated in ancient times for the blue pigments (woad) obtained from the partly dried leaves. Nowadays a weed of cornfields and naturalized on cliffs in the Severn Valley.

Mustard family *Brassicaceae (Cruciferae)*
Flowering time: June–Sept.
Height of growth: 15–80 cm
Dicotyledonous; Annual

Identification marks
Stem simple or branched from the middle, with hairs towards the base. Leaves lanceolate, the upper ones sessile with arrow-shaped base, the lower ones almost petiolate. Inflorescence elongating in fruit. Flowers golden yellow, approx. 0.5 cm wide. Fruits spherical, 2 mm across, with a network of fine wrinkles, on stalks projecting outwards at an angle.

Habitat and distribution
Arable fields, field paths and waste places. On dry warm loamy soils containing lime and nitrogen. Occurring as a casual.

Additional information
Other name: *Vogelia paniculata*. Similar field weeds (e.g. Gold-of-pleasure, *Camelina sativa*) have virtually been wiped out.

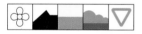

Biscutella laevigata

Yellow Whitlow-grass
Draba aizoides

Mustard family *Brassicaceae (Cruciferae)*
Flowering time: May–Nov.
Height of growth: 15–30 cm
Dicotyledonous; Perennial

Mustard family *Brassicaceae (Cruciferae)*
Flowering time: March–May
Height of growth: 5–15 cm
Dicotyledonous; Perennial

Identification marks
Stem erect, often branched towards top, leaves sparse. Basal leaves are long, margin entire or toothed, usually forming dense rosettes. Loose or dense panicles. Petals pale yellow, 0.5 cm long. Ovaries and fruit (silicula) made up of 2 flat, circular sections resembling spectacles.

Habitat and distribution
Not British. Dry stony turf, rocky ledges, Alpine scree, pine and mountain pine forests. Calcicolous; scattered in the Alps, otherwise rare.

Additional information
Falls into several ecological and geographical categories. Some forms grow on chalky ground and others on limestone moorland. A very variable species with different chromosomal races occupying alpine/lowland habitats. Numerous ssp.

Identification marks
Compact, semi-spherical rosettes of lanceolate leaves, barely 2 cm long, rigid, margin edged with stiff bristle-like hairs. Stems leafless. Corymbose racemes of golden-yellow flowers, relatively large, about 1 cm across. Fruits flat, elliptical, approx. 3 times longer than wide.

Habitat and distribution
In Britain only found on limestone in one locality in Glamorgan.

Alternate-leaved Golden Saxifrage
Chrysosplenium alternifolium

Opposite-leaved Golden Saxifrage
Chrysosplenium oppositifolium

Saxifrage family *Saxifragaceae*
Flowering time: April–July
Height of growth: 8–15 cm
Dicotyledonous; Perennial

Identification marks
Stem triangular, breaks easily. Several long-stalked, reniform basal leaves, 1–3 stem leaves similar to the lower ones, alternate. Inflorescence dichotomous, bracts almost sessile, yellow at the top. Flowers small, rich yellow in colour.

Habitat and distribution
Moist deciduous woods, streams, around springs, marshy areas, wet upland meadows. Local in most places but absent from western parts of England and Wales.

Additional information
Similar inflorescence: species from the Spurge genus (*Euphorbia*), milky juice.

Saxifrage family *Saxifragaceae*
Flowering time: April–July
Height of growth: 5–15 cm
Dicotyledonous; Perennial

Identification marks
Stem quadrangular, numerous leafy, creeping, above-ground stems, forming a dense flat turf. Flowering stems ascending. Stem leaves round, coarsely crenate, short petioles, opposite. Basal leaves similar, somewhat larger. Inflorescence with yellowish bracts and small rather pale yellow (also greenish-yellow) flowers.

Habitat and distribution
Cool woodland brooks, saturated rocks, wet ditches, wet areas round springs. Usually in the shade. Calcifugous. Up to about 1,100 m. Common except in southern and eastern parts.

Wild Mignonette
Reseda lutea

Tormentil
Potentilla erecta

Mignonette family *Resedaceae*
Flowering time: June–Aug.
Height of growth: 30–75 cm
Dicotyledonous; Perennial

Identification marks
Stem branched, ascending to erect. Upper leaves pinnate or bipinnate. Dense racemes of flowers. Flowers pale yellow; usually 4 large deeply lobed petals, plus 2 very small ones.

Habitat and distribution
Rubble heaps, paths, railway embankments, arable fields. Calcicolous. Requires loose often stony ground rich in nutrients. Northern England southwards.

Additional information
Similar: Weld, *R. luteola*, common, in similar habitats. Old dyeing plant. Only 4 petals, all leaves undivided.

Rose family *Rosaceae*
Flowering time: June–Sept.
Height of growth: 10–30 cm
Dicotyledonous; Perennial

Identification marks
Stem prostrate to erect. Stem leaves alternate, sessile, with 3 leaflets and 2 large leaflet-like stipules. Basal leaves comprise 3 leaflets, along with thin petioles, often already withered by flowering time. Flowers on pedicels, solitary in leaf axils, about 1 cm wide.

Habitat and distribution
Heaths, dry meadows, moorland. On light soils. Calcifugous; indicator of surface acidity and lack of nutrients. Common, up to over 1,000 m.

Additional information
Old medicinal plant with many scientific names: *P. tormentilla, Tormentilla erecta*.

Wood Spurge
Euphorbia amygdaloides

Euphorbia brittingeri

Spurge family *Euphorbiaceae*
Flowering time: March–May
Height of growth: 30–80 cm
Dicotyledonous; Perennial

Identification marks
Plant contains white milky juice (latex). Stem ascending to erect, in 2 conspicuous sections (2-year growth): lower part leathery with a shock of large ovoid-spatulate leaves often covered with fine hairs, upper part herbaceous, few leaves, many-branched inflorescence. Umbels with 5–10 main rays. Glands of cyathium half-moon-shaped: capsule finely spotted (check with magnifying glass). Bracts are fused together in pairs.

Habitat and distribution
In deciduous and mixed deciduous woodland on slightly damp loamy soil rich in mull and nutrients and usually calcareous. Common in the south, only local in the north, apparently absent from Scotland.

Spurge family *Euphorbiaceae*
Flowering time: May–June
Height of growth: 30–50 cm
Dicotyledonous; Perennial

Identification marks
Plant contains white milky juice (latex). Stem ascending or erect, several stems growing out of the base of the decayed stalks from the previous year (plant growth is therefore in clusters). Leaves long and obovoid, sessile, margin entire or finely toothed. Glands of cyathium are oval: capsule is thick and warty, 3–4 mm in diameter.

Habitat and distribution
Not British. Balks, meagre turf, sunny thickets. Also on pastures and waysides. Calcicolous. Prefers rather dry deep soils poor in nutrients. Only in southern parts of northern Europe. Scattered. Up to about 1,000 m.

Additional information
Similar: Sweet Spurge (see p.362). Synonym: *E. verrucosa*.

Broad-leaved Spurge
Euphorbia platyphyllos

Sun Spurge
Euphorbia helioscopia

Spurge family *Euphorbiaceae*
Flowering time: June–Oct.
Height of growth: 15–80 cm
Dicotyledonous; Annual

Identification marks
Plant contains white milky juice (latex). Stem erect or curved at the base. Leaves sessile, finely toothed at the apex: lower ones ovoid with blunt ends, the upper ones rather more lanceolate, pointed. Umbel with 5 (rarely 3) loosely branched main rays. Glands of the cyathium are broadly oval: capsule warty, 2–3 mm in diameter.
Habitat and distribution
Fields, gardens; also along paths. On warm, well-fertilized soil which is not too dry. Widespread in England, local in parts and rarer in the north.
Additional information
Similar: adjacent entry and also *E. stricta*, umbel usually has 2–5 rays, capsule is small, max. 2 mm diameter. In woods on limestone areas of the southern England/Wales border.

Spurge family *Euphorbiaceae*
Flowering time: May–Oct.
Height of growth: 10–50 cm
Dicotyledonous; Annual

Identification marks
Plant contains white milky juice (latex). Stem usually erect, simple, sometimes with a few branches at the base. Leaf base is wedge-shaped, apex spatulate and finely-toothed, upper leaves enlarged. Umbel usually has 5 main rays. Glands of cyathium are oval: capsule is finely spotted (check with magnifying glass).
Habitat and distribution
Cultivated ground and as a wayside weed. On loose soils rich in nutrients and minerals. Nitrogen indicator. Very common throughout.
Additional information
Similar, often in the same location: Petty-leaved Spurge (*E. peplus*, see p.362); rather calcifugous. Broad-leaved Spurge (see entry left).

Cypress Spurge
Euphorbia cyparissias

Dwarf Spurge
Euphorbia exigua

Spurge family *Euphorbiaceae*
Flowering time: May–Aug.
Height of growth: 10–30 cm
Dicotyledonous; Perennial

Identification marks
Plant contains white milky juice (latex). Stem erect, many leaves. Leaves narrow-linear, barely 2 mm wide, bluish-green. Umbel with 9–15 rays. Glands of cyathium half-moon-shaped: capsule finely spotted and rough (check with magnifying glass).

Habitat and distribution
Grassland, scrubby and open areas and waste places. On sunny warm limestone ground poor in nutrients. Rather scattered, frequently an escape from cultivation or casual.

Additional information:
Yellowish plants with deformed leaf growth have been attacked by uredo (red pustules on the leaf undersides).
Similar: Leafy Spurge (*E. esula*), leaves oblanceolate, approx. 3 mm wide. Very rare. The distribution is not fully known.

Spurge family *Euphorbiaceae*
Flowering time: June–Oct.
Height of growth: 5–30 cm
Dicotyledonous; Annual

Identification marks
Plant contains white milky juice (latex). Stem ascending to erect, often with many branches. Leaves linear, 1–4 mm, pointed, sessile. Umbel with 3 (–5) forked main rays. Glands of the cyathium are half-moon-shaped; capsule smooth.

Habitat and distribution
Arable fields. Prefers dry calcareous loamy soil rich in nutrients. Likes warmth. Common in England and Wales, also southern Scotland but becoming very rare in the north.

Additional information
Remotely similar: Sickle Spurge (*E. falcata*), fields; rare (Mediterranean plant). Leaves wider, lanceolate, bluish-green.

Common Evening-primrose
Oenothera biennis

Willowherb family *Onagraceae (Oenotheraceae)*
Flowering time: June–Sept.
Height of growth: 5–100 cm
Dicotyledonous; Biennial

Identification marks
Flowers large, over 3 cm wide; petals longer than stamens; sepals narrow, folded back. Spikes with many flowers. Stem erect, often unbranched. Leaves long, ovoid, stiff, toothed or entire.

Habitat and distribution
Railway embankments, rubble heaps, river banks, quarries, paths, waste places. Usually in dry soil. Once a widespread casual, now decreasing though naturalized in some places.

Additional information
Many similar types differing in minor characteristics (flower size, hairy growth, red flecks): Small-flowered Evening-primrose (*O. parviflora* = *muricata*), flowers smaller than 3 cm, stamens long.

Yellow Bird's-nest
Monotropa hypopitys

Wintergreen family *Purolaceae*
Flowering time: June–Aug.
Height of growth: 10–30 cm
Dicotyledonous; Perennial

Identification marks
Plant without any chlorophyll, pale yellow to whitish, more rarely reddish or tinged with brown. Leaves scale-like. Stem erect. Flowers in a terminal raceme, initially drooping. Flowers campanulate, 4 lobed, only terminal flower has 5 lobes, all 1–2 cm long, hanging slightly, erect in fruit.

Habitat and distribution
Coniferous and mixed woodland, more rarely in pure deciduous woods on poor soil. Needs acid soil rich in mull and well-moistened. Widespread but never common throughout England, in north and south but not central Wales and parts of Scotland.

Additional information
Two subspecies: Ssp. *hypopitys*, up to 11 flowers, dense inflorescence, usually covered with hairs. Ssp. *hypophegea*, 1–6 glabrous flowers.

Crosswort
Galium cruciata

Lady's Bedstraw
Galium verum

Madder family *Rubiaceae*
Flowering time: May–June
Height of growth: 15–70 cm
Dicotyledonous; Perennial

Madder family *Rubiaceae*
Flowering time: July–Aug.
Height of growth: 15–100 cm
Dicotyledonous; Perennial

Identification marks
Entire plant covered with short hairs. Stem quadrangular, slender, prostrate to erect. Leaves in whorls of 4, light green, ovoid, with 3 veins. Flowers in leaf axils, small, barely 3 mm wide.

Habitat and distribution
Woodland margins, thickets, hedges, waysides, more rarely on meadows near woodland. On well-moistened nitrogenous soil rich in humus and often low in lime. Throughout Britain except the extreme north.

Additional information
Previously attributed to a different genus under the name of *Cruciata laevipes*. Further synonym: *C. chersonensis*.

Identification marks
Stem bluntly 4-angled, ascending to erect. Leaves needle-shaped, single-veined, with short, sharp tips, margins curled up; in whorls of 8–12. Many-flowered terminal panicles. Flowers small, barely 3 mm wide, sweet-smelling.

Habitat and distribution
Grassland, hedges, dunes. Likes calcareous soils which are dry and warm in the summer. Very common throughout.

Additional information
The hybrid *G. x pomeranicum* (with Hedge Bedstraw – *G. mollugo*, see p.47) has pale yellow flowers.

Yellow Water-lily
Nuphar lutea

Marsh-marigold
Caltha palustris

Water-lily family *Nymphaeaceae*
Flowering time: June–Aug.
Height of growth: 0.5–2.5 m
Dicotyledonous; Perennial

Identification marks
Leaf-blades and flowers floating on the water surface, growing from rope-like stems. Flowers 4–6 cm wide. Leaves broadly ovoid with heart-shaped basal notch, 12–40 cm long; lateral veins forking near the margin, not joining together.

Habitat and distribution
Stagnant or slow-flowing water; likes cool water rich in nutrients and occasionally acid. Throughout Britain but rarer in northern Scotland.

Additional information
Similar: Least Water-lily (*N. pumila*), flowers 1–3 cm wide, leaves 4–14 cm long. Local, mainly in Scotland but also a few in England and Wales, in cold lakes low in nutrients. Leaves of White Water-lily (p.84) can be identified by lateral veins linking together along leaf margins.

Buttercup family *Ranunculaceae*
Flowering time: March–July
Height of growth: 15–50 cm
Dicotyledonous; Perennial

Identification marks
Leaves and flowers have glossy sheen. Stem hollow, prostrate to ascending, branched. Leaves reniform, finely toothed margin; the upper ones sessile with conspicuous herbaceous sheaths, the remainder petiolate; petioles grooved. Perianth is composed of only 5 segments (no separate petals and sepals); they measure up to 5 cm in diameter.

Habitat and distribution
Wet woodland and meadows, river banks, ditches, streams, areas round springs and reed beds. Likes soil which is rich in nutrients and saturated with ground water. Common throughout. Up to about 1,200 m.

Additional information
Many races; in particular the lowland type, and the type which grows in relatively hilly and mountainous areas.

Yellow Anemone
Anemone ranunculoides

Goldilocks Buttercup
Ranunculus auricomus

Buttercup family *Ranunculaceae*
Flowering time: April
Height of growth: 10–20 cm
Dicotyledonous; Perennial

Identification marks
Usually 2 (1–4) stalked flowers emerge from a single whorl of stem leaves. Stem otherwise bare of leaves, erect. Very few basal leaves which appear after flowering, petiolate, palmately compound (3 leaflets). Flowers 1–2 cm wide, layer of downy hairs on the outside.
Habitat and distribution
Naturalized in a few places in England.

Buttercup family *Ranunculaceae*
Flowering time: April–May
Height of growth: 10–40 cm
Dicotyledonous; Perennial

Identification marks
Stem erect, branched, few hairs. 2–6 basal leaves, petiolate, ranging in shape from undivided with a reniform outline to 5-lobed. Stem leaves conspicuously different from basal leaves: sessile and divided virtually to the base into narrow lobes. Few flowers, 1–2.5 cm across, occasionally with stunted petals or ones which wither rapidly.
Habitat and distribution
Deciduous woodland, mixed woodland, lowland forests, thickets, undergrowth along streams. Usually found on calcareous loamy soils saturated with ground water. Common throughout.

Corn Buttercup
Ranunculus arvensis

Meadow Buttercup
Ranunculus acris

Buttercup family Ranunculaceae
Flowering time: June–July
Height of growth: 15–60 cm
Dicotyledonous; Annual; Poisonous

Identification marks
Stem erect, branched. Lower leaves undivided, wedge-shaped, toothed – often dried up by the time the flowers blossom; stem leaves divided into narrow segments. Many sulphur-yellow flowers 0.5–1.5 cm in diameter, on hairy peduncles. Fruit conspicuously prickly.

Habitat and distribution
Especially in corn fields but also found on waste land. On loamy soil rich in nitrogen and preferably containing lime. Common in the south, rarer in the north, in some places heavily reduced presence as a result of the use of herbicides.

Buttercup family *Ranunculaceae*
Flowering time: May–July
Height of growth: 15–100 cm
Dicotyledonous; Perennial

Identification marks
Stem erect, branched. Lower leaves petiolate, palmately divided, lobes in turn divided or long toothed. Upper stem leaves sessile, less divided, segments narrow. Inflorescence cymose. Peduncles hairy, but not grooved. Calyx spreading but not reflexed.

Habitat and distribution
Meadows and pastures which are not too dry. Likes damp nitrogenous loamy soils. Very common; up to about 1,300 m.

Additional information
Similar: in addition to the species on the following page: Wood Buttercup (*R. nemorosus*), peduncles sparsely hairy but with longitudinal grooves. Not British. Light woodland, forest margins, mountain meadows; scattered. Several subspecies.

Bulbous Buttercup
Ranunculus bulbosus

Creeping Buttercup
Ranunculus repens

Buttercup family *Ranunculaceae*
Flowering time: May–July
Height of growth: 15–40 cm
Dicotyledonous; Perennial

Identification marks
Sepals folded back and lying close to the longitudinally grooved peduncle. Stem swollen at the base to form a bulb-like stem-tuber (just below the soil surface). Basal leaves have long petioles and are divided into 3 leaflets, the middle one with a conspicuous stalk. All leaflets may themselves be divided into 3s projecting at base of stem, but clinging towards top.
Habitat and distribution
Dry meadows, balks, path verges. Prefers rather poor warm loamy soil, usually calcareous. Throughout Britain, often very common becoming less so in the north.
Additional information
Similar: Hairy Buttercup (*R. sardous*). Covered with projecting hairs, no bulb; flowers pale yellow. Local, usually as a weed in fields.

Buttercup family *Ranunculaceae*
Flowering time: May–Aug.
Height of growth: 15–60 cm
Dicotyledonous; Perennial

Identification marks
Stem usually ascending, with creeping runners above the ground, often rooting at the nodes. Basal leaves divided into 3 segments each of which is further divided, middle lobe having a long, conspicuous stalk. Flowers solitary in the leaf axils, on peduncles. Sepals spread out. Peduncles have longitudinal grooves.
Habitat and distribution
Edges of banks, wet fields, gardens, path verges, damp meadows, woods. On moist heavy loamy soils. Nitrogen indicator. Common throughout. Often in masses.
Additional information
Very variable in hair covering, leaf form, flower size (and form) but these variations are apparently only habitat-related (not transmitted through reproduction).

Celery-leaved Crowfoot
Ranunculus sceleratus

Woolly Buttercup
Ranunculus lanuginosus

Buttercup family *Ranunculaceae*
Flowering time: May–Sept.
Height of growth: 20–60 cm
Dicotyledonous; Annual

Identification marks
Flowers barely 1 cm across, pale yellow; sepals reflexed but falling off rapidly, as long as or longer than the petals. Peduncles have longitudinal grooves. Stem ascending (also floating in water), hollow, many branches. Leaves somewhat fleshy, divided into 3 narrow lobes, sometimes themselves lobed and toothed. Fruiting head cylindric; 70–100 small fruits.

Habitat and distribution
River banks, ditches, mud, shallow pools. Found in soil which is at least occasionally flooded, and must be very wet and rich in nutrients. Rare. Throughout Britain but scattered and rarer in north Scotland. Tolerates salt.

Additional information
Many habitat-related types: from the stunted growth found on dry soils to the aquatic plant.

Buttercup family *Ranunculaceae*
Flowering time: May–July
Height of growth: 30–70 cm
Dicotyledonous; Perennial

Identification marks
Entire plant covered with dense projecting hairs. Stem erect, usually branched, hollow at the bottom, with yellow tufts. Basal leaves petiolate, divided into 5 lobes; lobes are wide, ovoid, toothed; stem leaves are similar but with short petioles or sessile. Flowers large, approx. 3–4 cm wide. Sepals spread out; peduncles not longitudinally grooved.

Habitat and distribution
Not British. Damp shady deciduous and mixed woodland, thickets, undergrowth along streams. On loamy soils saturated with ground water and rich in nitrogen, lime and mull. Scattered. Eastern parts of C. and S. Europe.

Additional information
Similar: Wood Buttercup (*R. nemorosus*), more delicate peduncle with longitudinal grooves (p.134).

Great Spearwort
Ranunculus lingua

Lesser Spearwort
Ranunculus flammula

Buttercup family *Ranunculaceae*
Flowering time: June–Sept.
Height of growth: 50–120 cm
Dicotyledonous; Perennial

Identification marks
Leaves undivided, linear-lanceolate, the basal ones withering before flowering. Stem creeping then becoming erect, usually many branches towards the top, hollow. Usually large number of flowers, with stalks, 2.5–4 cm wide.

Habitat and distribution
In reed beds in stagnant or slow-flowing waters, also in ditches, marshes and fens. On muddy ground which is occasionally flooded and is rich in nutrients. Local but widespread.

Additional information
This species grows both on banks and in shallow water up to over 50 cm depth and also grows in different forms depending on the varying conditions.

Buttercup family *Ranunculaceae*
Flowering time: May–Sept.
Height of growth: 8–50 cm
Dicotyledonous; Perennial

Identification marks
Stem fairly thick; prostrate, ascending or erect. Leaves undivided, narrow-lanceolate to ovate, sometimes spoon-shaped with long petioles. Usually numerous flowers, on long peduncles rarely over 2 cm wide.

Habitat and distribution
Banks, streams, ponds, ditches, damp meadows, moors, beds of sedge, wet pathways, more rarely in reed beds. Found in wet, muddy or loamy soil, usually acid. Common.

Additional information
Similar: Creeping Spearwort (*R. reptans*), flowers less than 1 cm, all leaves petiolate in tufts at each point where the creeping stem roots. Prefers soil which is richer in nutrients and more alkaline, rarer.

Orpine
Sedum telephium

Biting Stonecrop
Sedum acre

Stonecrop family *Crassulaceae*
Flowering time: uly–Sept.
Height of growth: 20–60 cm
Dicotyledonous; Perennial

Stonecrop family *Crassulaceae*
Flowering time: June–July
Height of growth: 2–10 cm
Dicotyledonous; Perennial

Identification marks
Leaves ovoid, smooth, flat but fleshy. Stem erect. Many flowers in dense cymes at top of stem.

Habitat and distribution
In light woodland and hedgerows, often on banks. Likes moderately dry, stony ground rich in nutrients. Scattered and rather local in most parts, rare in northern Scotland; also growing as a garden escape (old ornamental plant).

Additional information
Various colour types: Ssp. *maximum*, usually with greenish-yellow (more rarely light reddish) flowers; ssp. *telephium*, flowers usually reddish or lilac (more rarely greenish); ssp. *fabaria*, flowers reddish or lilac; ssp. *ruprechtii*, flowers white.

Identification marks
Leaves thick and fleshy, upper sides flat, with ovoid outline, approx. 0–5 cm long. Stem prostrate or ascending. Inflorescence with few flowers. Petals pointed, 0.5–1 cm long.

Habitat and distribution
Walls, gravel paths, sand dunes, rock fissures, scree, stony river beds, sandy or stony turf. Found on dry, shallow and often stony calcareous ground.

Additional information
Also known as Wall-pepper. Similar: Tasteless Stonecrop, *S. sexangulare* (*S. mite*, *S. boloniense*), flowers smaller. Naturalized in some places in England and Wales.

Wood Avens
Geum urbanum

Creeping Avens
Geum reptans

Rose family *Rosaceae*
Flowering time: June–Aug.
Height of growth: 20–60 cm
Dicotyledonous; Perennial

Identification marks
Stem erect, usually unbranched. Basal rosette of 3–7 pinnate leaves with unequal leaflets and short petioles. Stem leaves 3–5 leaflets or entire. Cymes loose and 4-flowered. Sepals about as long as petals, plus 5 shorter epicalyx segments. Fruit has hooked awn.

Habitat and distribution
Deciduous and mixed woodland, woodland paths and margins, thickets, balks, hedgerows and shady places. Likes well-moistened nitrogenous soil. Common.

Additional information
Old medicinal plant. Also called Herb Bennet.

Rose family *Rosaceae*
Flowering time: July–Aug.
Height of growth: 5–15 cm
Dicotyledonous; Perennial

Identification marks
Stem mostly erect, flower single. Flowers up to 4 cm wide. Basal leaves form rosette. Leaves pinnate, leaflets deeply cleft, getting larger towards the leaf apex, terminal leaflet is not however exceptionally large. Produces long twisted stolons. Fruits numerous; all have long tufted, hairy style.

Habitat and distribution
Not British. Only found in the High Alps between 2,000 and 2,400 m on damp unstable scree and moraines. Rare.

Additional information
Similar: Alpine Avens (*G. montanum*), without stolons, terminal leaflet markedly larger. Alpine turfs and heaths (1,500–2,300 m). Not British.

Golden Cinquefoil
Potentilla aurea

Alpine Cinquefoil
Potentilla crantzii

Rose family *Rosaceae*
Flowering time: June–Aug.
Height of growth: 5–20 cm
Dicotyledonous; Perennial

Identification marks
Stem usually curved at the base rising to erect; branched. Flowers in panicles, up to 2 cm across, long peduncles. Leaves palmately 5 lobed – stem leaves sometimes with only 3 lobes. Leaflets wedge-shaped to narrow ovoid, toothed at the apex, margin covered with silky hairs.

Habitat and distribution
Not British. Meagre Alpine turf, pastures, snowy valleys, heaths lying between 1,000 and 2,400 m. Likes acid soil low in lime and not too dry. Only the Alps and Black Forest. Scattered.

Additional information
Similar Alpine Cinquefoils: Alpine Cinquefoil (*P. crantzii*), see below. *P. brauneana*, leaves have only 3 lobes; stony ground, pastures; calcicolous. Not British.

Rose family *Rosaceae*
Flowering time: March–May
Height of growth: 5–25 cm
Dicotyledonous; Perennial

Identification marks
Stem creeping to ascending. Leaves digitate, 5–7 lobes; leaflets toothed, leaf margin does not have silky sheen. Cyme with few flowers. Flowers 1–2 cm across; petals deeply notched, not overlapping at the edge.

Habitat and distribution
Sunny balks, dry pastures, sandy slopes. Calcicolous, likes warmth. Very local, usually on high ground. Northern England, north Wales, Scotland.

Additional information
A very variable species, further confused by the presence of hybrids between it and other cinquefoils.

Creeping Cinquefoil
Potentilla reptans

Silverweed
Potentilla anserina

Rose family *Rosaceae*
Flowering time: June–Aug.
Height of growth: 30–100 cm
Dicotyledonous; Perennial

Identification marks
Stem prostrate, creeping, stolon-like, rooting at the nodes. Leaves petiolate, palmately divided, usually 5 leaflets (more rarely 3 or 7). Leaflets obovoid or oblong, toothed, sparsely hairy on both sides. Flowers solitary in leaf axila, 1.5–2.5 cm wide.

Habitat and distribution
Relatively damp meadows, hedgerows, waste places, also fields and gardens. Likes nitrogenous soil and warmth. Common.

Additional information
Similar: Trailing Tormentil (*P. anglica = procumbens*), scattered but local, rare in north Scotland. Flowers max. 1.8 cm wide, often only 4 petals. This fertile species is derived from the sterile hybrid of Tormentil (see p.126) and Creeping Cinquefoil.

Rose family *Rosaceae*
Flowering time: June–Aug.
Height of growth: 15–50 cm
Dicotyledonous; Perennial

Identification marks
Stem creeping to erect. Leaves pinnate; 7–12 pairs of oblong, deeply serrate main leaflets alternating with smaller leaflets, undersides covered with silky hairs. Flowers solitary, on long peduncles, golden-yellow, approx. 2 cm wide.

Habitat and distribution
Path verges, village commons, railway embankments, grassy beaches, fallow land, damp pastures and waste places. Likes warmth and nitrogen, tolerates salt. Prefers compact loamy soils. Common, reaching about 450 m.

Agrimony
Agrimonia eupatoria

Upright Yellow Oxalis
Oxalis europaea

Rose family *Rosaceae*
Flowering time: June–Aug.
Height of growth: 30–60 cm
Dicotyledonous; Perennial

Identification marks
Many small flowers approx. 1 cm across in spike-like racemes. Stem erect, covered with rough hairs. Leaves pinnate, leaflets serrate; smaller pinnate leaflets growing between the large ones. Fruit with hooked spines.

Habitat and distribution
Path balks, sunny slopes, pastures, woodland margins, hedgerows. Likes calcareous humus soil, not too dry but low in nutrients. Likes warmth. Common in most parts up to about 500 m but rare in northern Scotland.

Additional information
The robust ssp. *grandis* from Europe is larger and hairier than ssp. *eupatoria*.

Wood-sorrel family *Oxalidaceae*
Flowering time: June–Sept.
Height of growth: 10–30 cm
Dicotyledonous; Annual – Perennial

Identification marks
Stem erect, not rooting at the nodes, hairy. Leaves petiolate, trifoliate like clover, stipules absent. Inflorescence cymose, with 1–6 flowers. Capsules angular, on erect pedicels.

Habitat and distribution
Not British. Fields, gardens, paths, rubble heaps. Usually on soil which is low in lime and contains no nitrogen. Frequent as a weed in most parts of Europe except the far north and south.

Additional information
Synonym: *O. stricta*. Similar: Procumbent Yellow Oxalis (*O. corniculata*), stem creeping, rooting at the nodes, often red. Small stipules at base of leafstalk. Scattered, sometimes common. Gardens, fields, tarmac roads, pavements. Both species are introduced synanthropic plants.

Common Rock-rose
Helianthemum nummularium

Trailing
St John's-wort
Hypericum humifusum

Rock-rose family *Cistaceae*
Flowering time: June–Sept.
Height of growth: 5–30 cm
Dicotyledonous; Perennial

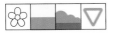

St John's-wort family *Hypericaceae*
Flowering time: June–Sept.
Height of growth: 5–20 cm
Dicotyledonous; Biennial – Perennial

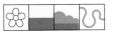

Identification marks
Stem prostrate to erect, woody at the base. Leaves opposite, oblong or oval, margin entire. Unilateral cymes with few (1–12) flowers. Flowers usually about 2–3 cm wide; 3 larger sepals, 2 very small inner ones.

Habitat and distribution
Poor turf, pastures, light thickets. Likes warmth. Prefers stony usually calcareous ground. Common in most parts but absent from the extreme south west of England, Isle of Man and north west Scotland.

Additional information
Many subspecies: Ssp. *nummularius* (leaf underside has grey felt covering) is main form in lower-lying areas. Ssp. *grandiflorum* (curly hairs) and ssp. *glabrum* (virtually glabrous): Alpine varieties up to 2,300 m have large flowers (3–4 cm).

Identification marks
Stem slender, branched, creeping, curved upwards at the ends; usually has a narrow rib along 2 sides of stem. Leaves opposite, oblong-ovoid. Inflorescence with few flowers. Flowers 1–1.5 cm wide. Plant glabrous.

Habitat and distribution
Dry moorland, heaths, open woodland. Likes open soil low in lime. Scattered.

Hairy St John's-wort
Hypericum hirsutum

Pale St John's-wort
Hypericum montanum

St John's-wort family *Hypericaceae*
Flowering time: July–Aug.
Height of growth: 40–100 cm
Dicotyledonous; Perennial

Identification marks
Entire plant has a thick covering of short hairs. Stem twisting at base rising to erect, round. Leaves opposite, margin entire, ovoid, blunt, virtually sessile; leaf surface has translucent spots but no black glands. Inflorescence loose, many flowered. Flowers around 1.5 cm wide. Sepals have black glands around margin, petals have fewer.

Habitat and distribution
Somewhat wet open woodland, thickets, grassy areas. Found on calcareous soils rich in nitrogen. Scattered.

St John's-wort family *Hypericaceae*
Flowering time: June–Aug.
Height of growth: 40–80 cm
Dicotyledonous; Perennial

Identification marks
Plant slightly hairy. Stem erect, round. Leaves opposite, glabrous above, thinly hairy beneath, sessile, half clasping the stem, with black dots round the edge beneath; only the upper leaves have translucent spots on the leaf blade. Inflorescence dense. Flowers 1–1.5 cm wide. Sepals have black marginal glands. Approx. 40–60 stamens in 3 bundles.

Habitat and distribution
Dry thickets and woodland. Found on ground which is rich in nutrients and usually calcareous. Scattered.

Additional information
In appearance very close to the previous entry; can however be distinguished clearly by the leaves being glabrous above and by the dense, short inflorescence.

Imperforate St John's-wort
Hypericum maculatum

Square-stalked St John's-wort
Hypericum tetrapterum

St John's-wort family *Hypericaceae*
Flowering time: June–Aug.
Height of growth: 20–60 cm
Dicotyledonous; Perennial

Identification marks
Plant glabrous. Stem erect, with 4 narrow longitudinal ribs. Leaves opposite, sessile, barely or not at all translucently spotted, with black marginal glands. Flowers approx. 2 cm wide. Sepals and petals often have light-coloured or black spots. Stamens in 3 bundles.

Habitat and distribution
Hedgerows and woodland margins. Found on slightly acid ground which is at least occasionally wet. Scattered throughout most of Britain but rather local.

Additional information
H. x desetangsii is similar but the stem has 2 distinct ribs and 2 faint ones. In similar habitats but rare.

St John's-wort family *Hypericaceae*
Flowering time: June–Sept.
Height of growth: 30–70 cm
Dicotyledonous; Perennial

Identification marks
Plant glabrous. Stem erect, branched, with 4 wide wing-like longitudinal ribs. Leaves opposite, sessile, half-clasping the stem; leaf blade fine and covered with translucent spots. Additionally the leaves, ribs, petals and sepals have black glandular spots. Inflorescence compact with many flowers. Flowers approx. 1 cm across. Stamens in 3 bundles.

Habitat and distribution
Wet meadows, grassy areas around springs, ditches, lakes and ponds. On wet, occasionally flooded soils rich in nutrients and preferably calcareous. Throughout Britain except northern Scotland.

Additional information
Synonym: *H. acutum.*

Perforate
St John's-wort
Hypericum perforatum

Slender
St John's-wort
Hypericum pulchrum

St John's-wort family *Hypericaceae*
Flowering time: June–Sept.
Height of growth: 30–90 cm
Dicotyledonous; Perennial

Identification marks
Plant glabrous. Stem erect, often branched, round, with 2 longitudinal ribs. Leaves opposite, sessile, narrowly oval, spotted with translucent glands. Stamens in 3 bundles.
Habitat and distribution
Open woodland, hedges, clearings, poor meadowland and pastures. On various different soils; usually calcareous, low in nitrogen, moderately dry. Common throughout most areas.
Additional information
Old medicinal magic plant (petals produce a reddish colour when crushed).

St John's-wort family *Hypericaceae*
Flowering time: June–Aug.
Height of growth: 30–60 cm
Dicotyledonous; Perennial

Identification marks
Plant glabrous. Stem erect or ascending, round. Leaves opposite, sessile, bluntly triangular to heart- shaped, half clasping the stem, translucently spotted. Inflorescence few flowered. Flowers approx. 1.5 cm wide. Sepals and petals have black marginal glands. Stamens in 3 bundles. Stem, leaves and flowers often tinged with red.
Habitat and distribution
Dry open woodland, glades and grassy areas. On acid sandy or loamy soil. Calcifugous. Indicator of poor soil. Avoids wet habitats. Scattered.

Sickle-leaved Hare's-ear
Bupleurum falcatum

Umbellifer family *Apiaceae (Umbelliferae)*
Flowering time: July–Oct.
Height of growth: 50–130 cm
Dicotyledonous; Perennial

Identification marks
Stem erect, hollow. Leaves leathery, usually very narrow, the upper ones sickle-shaped, margin entire. Flowers small, in compound umbels containing 4–10 rays; 2–5 bracts, often unequal, 4–5 large bracteoles which are shorter than the pedicels.

Habitat and distribution
Hedgerows and waste places. Found in loose calcareous soils which are low in nitrogen and where the location is warm in the summer. Only in a few places in south-east England.

Additional information
The Hare's-ear genus is the only one of the Umbelliferae found in Britain to have undivided leaves.

Wild Parsnip
Pastinaca sativa

Umbellifer family *Apiaceae (Umbelliferae)*
Flowering time: July–Aug.
Height of growth: 30–150 cm
Dicotyledonous; Biennial

Identification marks
Stem erect, branched at the top, grooved, hairy. Leaves coarsely simple pinnate. Compound umbel with 5–15 rays. Bracts and bracteoles absent, occasionally 1–2 but falling quickly. Flowers barely 2 mm wide. Petals often curled up, golden-yellow.

Habitat and distribution
Rubble heaps, path and road verges, fields, railway embankments. Likes nitrogenous deep loamy soil, especially on chalk and limestone. Found throughout England and Wales, sometimes locally common. In Scotland only an escape.

Additional information
Wild: Ssp. *urens*, grey hairs, stem finely grooved, S., C., E. Europe. Ssp. *sylvestris* (grooved, hairy), C., W., Europe. Wild as an escape. Ssp. *sativa*, almost glabrous, stem grooved.

Primrose
Primula vulgaris

Bear's-ear
Primula auricula

Primrose family *Primulaceae*
Flowering time: Dec.–May
Height of growth: 5–15 cm
Dicotyledonous; Perennial

Identification marks
Basal leaves oblong to obovoid gradually narrowed at base, occasionally lasting through winter, sometimes killed by frost. Young leaves begin to develop during flowering time; wrinkled, uppersides glabrous, underneath light covering of short hairs. Flowers on hairy pedicels arising from basal rosette. Calyx cleft almost to middle. Corolla 2–3 (+) cm wide, with notched lobes.

Habitat and distribution
Woodland, thickets, meadows, balks. On damp soils rich in nutrients and mull but rather low in lime. Common, less so in the north.

Additional information
Synonyms: *P. acaulis*, meaning without stem. Common ornamental plant (flowers white, yellow, red, brown, blue).

Primrose family *Primulaceae*
Flowering time: April–July
Height of growth: 5–30 cm
Dicotyledonous; Perennial

Identification marks
Rosette made up of slightly fleshy leaves with cartilaginous margins, hairless and mealy. Many-flowered often one-sided umbel on robust scape. Corolla rather campanulate, approx. 1.5 cm wide, the 5 lobes only slightly notched.

Habitat and distribution
Not British. Rocks, stony turf, marshy meadows. Found on very wet ground often low in humus. Calcicolous. In the Alps scattered up to 2,400 m, rare in the foreland; found northwards as far as the Schwäbisches Alb and the Black Forest.

Additional information
Similar: *P. hortensis*. Many-coloured ornamental plant based on hybrid of Auricula and Hairy Primrose *(P. hirsuta)* = *P. x pubescens*. Corolla margin flat, lobes deeply crenate.

Oxlip
Primula elatior

Cowslip
Primula veris

Primrose family *Primulaceae*
Flowering time: April–May
Height of growth: 10–30 cm
Dicotyledonous; Perennial

Identification marks
Umbel on tall scape, often one-sided. Flowers sulphur yellow; corolla margin flat, around 2 cm wide; calyx narrow, cylindrical, with green edges. Rosette leaves oblong-ovoid, wrinkled, crenate, with short hairs.

Habitat and distribution
Woodland where it occurs on chalky boulder clay. Locally common in small areas. South Midlands eastwards to East Anglia.

Primrose family *Primulaceae*
Flowering time: April–May
Height of growth: 10–30 cm
Dicotyledonous; Perennial

Identification marks
One-sided umbel on straight scape. Flowers golden-yellow, with orange flecks on the inside; margin of the corolla approx. 1.5 cm wide, campanulate; calyx inflated, light greyish-yellow. Rosette leaves oblong-ovoid, wrinkled, notched.

Habitat and distribution
Balks, meadow slopes, poor turf, thickets, open woodland. Found on soil which is preferably calcareous, not too moist and which contains nutrients. Fairly common, rarer in the north, and absent from some parts of Scotland.

Tufted Loosestrife
Lysimachia thyrsiflora

Yellow Loosestrife
Lysimachia vulgaris

Primrose family *Primulaceae*
Flowering time: May–June
Height of growth: 30–70 cm
Dicotyledonous; Perennial

Identification marks
Stem erect. Leaves opposite and decussate, lanceolate to linear-lanceolate, margin entire, sessile. Many dark gland spots on the leaf blade. Flowers in compressed cylindrical racemes in leaf axils, 3–5 mm wide, occasionally 6 in a raceme. Racemes are stalked but still not as long as the leaves.

Habitat and distribution
In marshes, canals and ditches with stagnant or slow-flowing waters. Found in muddy acid ground, not very rich in nutrients and frequently flooded. Rare. Scattered in central England and Scotland and parts of the south.

Additional information
Has been placed in a genus on its own: *Naumburgia*.

Primrose family *Primulacae*
Flowering time: July–Aug.
Height of growth: 60–150 cm
Dicotyledonous; Perennial

Identification marks
Stem erect, slightly angular. Leaves opposite or in whorls of 3–4, large, ovoid, up to 14 cm long with orange or black glands. Terminal panicles with leaves underneath. Flowers 1–2 cm wide; corolla lobes slightly obtuse, smooth margins, calyx lobes with orange margin which is glandular-hairy.

Habitat and distribution
Ditches, river banks, marshes, wet land with bushy growth. On wet peaty ground. Scattered throughout, rarer or absent in the north.

Additional information
Similar: Dotted Loosestrife (*L. punctata*). Flowers 3–3.5 cm wide, corolla lobes pointed, entire margin glandular-hairy; calyx lobes green, margin glandular-hairy. Ornamental plant, sometimes found wild.

Creeping Jenny
Lysimachia nummularia

Yellow Pimpernel
Lysimachia nemorum

Primrose family *Primulaceae*
Flowering time: June–Aug.
Height of growth: 5–60 cm
Dicotyledonous; Perennial

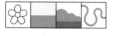

Identification marks
Stem far-creeping, virtually unbranched, occasionally ascending at the end. Leaves opposite, rounded, 1–3 cm long, short petioles. Flowers on stout pedicels, solitary, axillary; corolla approx. 1.5–2.5 cm across, sometimes spotted with red inside. Calyx lobes pointed. Plant usually glabrous.

Habitat and distribution
Meadows, ditches, waysides, fields, hedgerows. Found on heavy (often compressed) damp loamy soil rich in nitrogen. Reasonably tolerant of drought. Scattered throughout but rare in the north.

Additional information
Varied in the leaf form, number of flowers, scent and length of peduncle. Is occasionally grown in gardens.

Primrose family *Primulaceae*
Flowering time: May–Sept.
Height of growth: 10–40 cm
Dicotyledonous; Perennial

Identification marks
Stem ascending, hardly branched, rooting at the nodes lower down the plant. Leaves opposite and decussate, ovoid, up to 4 cm long, margin entire, on short petioles. Leaf blade densely covered with translucent spots. Flowers solitary in leaf axils, on very slender pedicels. Corolla approx. 1 cm across, lobes ovoid, obtuse, margin entire or toothed.

Habitat and distribution
Woods, thickets and hedgerows. On soil which is rich in nutrients but low in lime. Intolerant of drought. Less common in dry areas, otherwise common.

Great Yellow Gentian
Gentiana lutea

Smooth Honeywort
Cerinthe glabra

Gentian family *Gentianaceae*
Flowering time: July–Sept.
Height of growth: 30–60 cm
Dicotyledonous; Perennial

Identification marks
Stem robust, erect. Leaves bluish-green, opposite, broadly ovoid, with pronounced curved venation. Clusters of 3–10 flowers in axils of shell-shaped bracts, arranged in several stages above one another at end of stem. Flowers deeply cleft, 5–6 lobes.

Habitat and distribution
Not British. Balks, mountain meadows, Alpine mats, open mountain forests. Scattered in the Alps, otherwise very rare. Calcicolous.

Additional information
Similar in non-flowering state: False Helleborine (*Veratrum album*, see p.78), leaves alternate; also related Alpine gentians, all very rare with long ovoid-lanceolate leaves: Purple Gentian, Brown Gentian and Spotted Gentian (*G. purpurea, pannonica, punctata*).

Borage family *Boraginaceae*
Flowering time: May–July
Height of growth: 30–50 cm
Dicotyledonous; Perennial

Identification marks
Entire plant glabrous, with bluish tinge. Flowers in compact cylindrical clusters drooping at the end of the peduncle. Corolla campanulate-cylindrical, with 5 small somewhat reflexed lobes. Red ring or red spots in the throat of the tubular-shaped corolla. Leaves sessile with heart-shaped base.

Habitat and distribution
Not British. Pastures, Alpine mats, weedy places, scrub along brooks. Found on well-moistened ground rich in nitrogen and preferably calcareous. Rare; only in the (western) Alpine region and foreland. Occasionally planted in gardens.

Common Comfrey
Symphytum officinale

Henbane
Hyoscyamus niger

Borage family *Boraginaceae*
Flowering time: May–June
Height of growth: 30–120 cm
Dicotyledonous; Perennial

Identification marks
Entire plant covered with rough hairs. Leaves rather narrowly ovoid, distinctly decurrent. Flowers small campanulate, drooping, in scorpioidal cymes.

Habitat and distribution
On damp to wet ground, always rich in nutrients. In wet meadows, on river banks and in ditches. Found throughout Britain but less so in the north.

Nightshade family *Solanaceae*
Flowering time: June–Aug.
Height of growth: 30–80 cm
Dicotyledonous; Annual – Biennial

Identification marks
Plant covered with sticky hairs. Stem erect, often branched. Leaves oblong to ovoid, entire or with a few large teeth, lower ones petiolate, upper ones sessile and clasping the stem. Corolla funnel-shaped, 5 lobes, dirty yellow with network of violet-coloured veins.

Habitat and distribution
Rubble heaps, paths, walls, also in fields. Likes sunny warm loamy soil consistently rich in nitrogen and not too dry. Scattered and local throughout, in some areas only as a casual.

Dark Mullein
Verbascum nigrum

Aaron's Rod
Verbascum thapsus

Figwort family *Scrophulariaceae*
Flowering time: June–Oct.
Height of growth: 50–120 cm
Dicotyledonous; Biennial

Identification marks
Stamens have violet-coloured woolly hairs. Flowers about 1.5–2.5 cm wide; long terminal racemes with axillary short-stalked clusters of 5–10 flowers. Stem erect, little branching. Leaves elongate-ovoid, crenate, not decurrent, hairy on undersides.

Habitat and distribution
Wayside places and open areas. Found on slightly damp nitrogenous soils not always rich in lime but containing abundant nutrients. Southern England, where it is common, the Midlands and Wales, sometimes found further north.

Additional information
Similar: Moth Mullein (*V. blattaria*), rare; flowers solitary in the axils of bracts, leaves not hairy. White Mullein (*V. lychnitis*, p.75), corolla smaller, stamens covered with white wool.

Figwort family *Scrophulariaceae*
Flowering time: June–Aug.
Height of growth: 30–200 cm
Dicotyledonous; Biennial

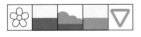

Identification marks
Stamens covered with white wool. Flowers 1.5–3 cm wide, in clusters, forming a long dense raceme. Stem erect, usually simple. Leaves oblong, slightly crenate, decurrent as far as the next leaf (stem therefore looks winged). Covered with dense whitish or greyish felt-like hairs.

Habitat and distribution
Pathways, railway embankments, rubble heaps, balks, heaths, forest margins. Likes warmth. Found on loose shallow often somewhat stony ground. Nitrogen indicator. Common in most areas, rarer in northern Scotland.

Additional information
Similar to the species described in the adjacent entry, which has larger flowers.

Large-flowered Mullein
Verbascum densiflorum

Yellow Bellflower
Campanula thyrsoides

Figwort family *Scrophulariaceae*
Flowering time: July–Sept.
Height of growth: 30–200 cm
Dicotyledonous; Biennial

Identification marks
Stamens covered with white wool, flowers 3–4 cm wide. Long dense raceme. Leaves narrowly ovate, strongly decurrent. Thick greyish-yellow felt-like covering of hairs.

Habitat and distribution
Wayside places, clearings, sometimes in fields. Likes warmth. Likes dry preferably calcareous soil rich in nitrogen. A rather rare casual.

Additional information
Synonym: *V. thapsiforme*. Similar: *V. phlomoides*, leaves not or only slightly decurrent (not as far as the next leaf). Rare casual.

Bellflower family *Campanulaceae*
Flowering time: July–Sept.
Height of growth: 10–40 cm
Dicotyledonous; Biennial

Identification marks
Flowers approx. 2 cm long, narrow-campanulate. Dense terminal conical-shaped spike. Stem erect, unbranched, many leaves. Leaves narrow, lanceolate, pointed; basal ones blunt, all sessile. Entire plant covered with dense layer of long bristle-like hairs (flowers rather more woolly-haired).

Habitat and distribution
Not British. Found exclusively in the Alps upwards of approx. 1,500 m on sunny calcareous and often stony ground which contains nutrients and is not too dry. Scattered (but very conspicuous). Only rarely washed down into the valley.

German Asphodel
Tofieldia calyculata

Bog Asphodel
Narthecium ossifragum

Lily family *Liliaceae*
Flowering time: May–July
Height of growth: 10–30 cm
Monocotyledonous; Perennial

Identification marks
Flowers 6-merous on short pedicels in terminal racemes 3–8 cm long. Stem erect, few leaves. Basal leaves grass-like, arranged in two rows, much shorter than the stem. The epicalyx at the base of the flower is 3-lobed and scale-shaped (in addition to the lanceolate bract on the main stem) and is characteristic of this species.

Habitat and distribution
Moors, marshy meadowland, damp poor turf, saturated rocks. Calcicolous. Dislikes fertilizer. Very rare, absent in the north.

Additional information
Similar: Scottish Asphodel, *T. pusilla* (= *palustris*). Local from northern England to Scotland. Does not have an epicalyx, flowers more whitish-green.

Lily family *Liliaceae*
Flowering time: July–Sept.
Height of growth: 10–40 cm
Monocotyledonous; Perennial

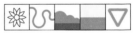

Identification marks
Flowers 6-merous in terminal raceme 5–8 cm long, outer flowers greenish. Stem erect, leafy. Leaves grass-like, basal ones arranged in 2 rows, approx. as long as stem and much longer than stem leaves. Young anthers orange; stamens have dense covering of woolly hairs. Stem and petals deep orange after flowering.

Habitat and distribution
High moorland, wet heaths, and mountains. Wet peaty soils low in nutrients. Calcifugous. Common where conditions are suitable, rarer in S. E. and absent from some parts of the Midlands and E. England.

Additional information
Previously classified under the same genus as the previous entry or as *Anthericum* (p.79) (*Tofieldia ossifraga* or *Anthericum ossifragum*).

Yellow Star-of-Bethlehem
Gagea lutea

Yellow Iris
Iris pseudacorus

Lily family *Liliaceae*
Flowering time: March–May
Height of growth: 10–25 cm
Monocotyledonous; Perennial

Identification marks
Single grass-like basal leaf 6–12 mm across. Umbel with 1–5 flowers which emerges from between 2 narrow bracts. Stem leafless. Pedicels glabrous; petals 1–1.5 cm long, with green stripes on the outside.

Habitat and distribution
Damp mixed deciduous woodland and grassland. Likes calcareous soil which is acid with humus, close to ground water and always rich in nutrients. Local in most parts, sometimes rare. Mainly in northern and central England.

Additional information
Similar: *G. pratensis*, not British, grassy areas, fields, rare. Basal leaf 3–8 mm wide, keeled. *G. villosa = arvensis*, not British, fields, rare; 2 basal leaves, 1–2 mm wide.

Iris family *Iridaceae*
Flowering time: May–July
Height of growth: 40–150 cm
Monocotyledonous; Perennial

Identification marks
3 outer petals ovoid, drooping slightly, without a beard of hairs; 3 inner petals smaller, spatulate, erect; stigmas petalloid. Stem erect, robust, leaves almost sheathed. Basal leaves stiff, sword-shaped, up to 3 cm wide, approximately as long as the stem. Capsule large, triangular.

Habitat and distribution
Marshy areas; also found in half-shade in damp undergrowth. Likes marshy ground rich in nutrients. Found wherever the right conditions occur.

Additional information
The yellow-flowered ornamental plants from this genus which are sometimes found growing wild are usually on drier ground and they all have a beard of hair on the inside of the outer petals.

Lesser Celandine
Ranunculus ficaria

Yellow Pheasant's-eye
Adonis vernalis

Buttercup family *Ranunculaceae*
Flowering time: March–May
Height of growth: 5–15 cm
Dicotyledonous; Perennial

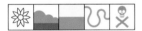

Identification marks
Stem hollow, prostrate to ascending. Leaves alternate, heart to kidney-shaped, crenate, the upper ones smaller; all have relatively long petioles, glossy. There are sometimes bulbils in the leaf axils. Flowers solitary, with long pedicels; 8–12 petals.

Habitat and distribution
Damp deciduous woodland and thickets, spring areas, river banks, meadows. Found on damp loamy soil rich in nutrients. Common throughout. Up to around 800 m.

Additional information
Falls into several sub-species which are not as yet completely defined. Synonym: *Ficaria verna*.

Buttercup family *Ranunculaceae*
Flowering time: April–May
Height of growth: 10–60 cm
Dicotyledonous; Perennial

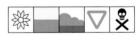

Identification marks
Flowers usually solitary, terminal, 4–8 cm wide, with 10–20 petals. Stem erect, sheathed scale-like leaves at the base, upper leaves 2–4 pinnately divided and split into many narrow lobes. As well as the peduncles there are also stalks bearing only leaves.

Habitat and distribution
Cultivated in Britain. Sunny balks, dry turf, bushy slopes, pine forests. On loose sandy calcareous soils. Very rare and becoming progressively rarer. Only found in lower-lying areas (up to 500 m).

Additional information
Like most early flowering plants this one may also flower again in autumn.

Globe-flower
Trollius europaeus

Wild Mignonette
Reseda lutea

Buttercup family *Ranunculaceae*
Flowering time: June–Aug.
Height of growth: 10–60 cm
Dicotyledonous; Perennial

Identification marks
Flower is a closed sphere up to 3 cm in diameter; 6–15 petals. Stem erect, usually only branched at the bottom; flowers solitary and terminal. Leaves palmate, 3–5 lobes. Basal leaves have long petioles.

Habitat and distribution
Damp meadows and woods in mountains. Likes a loamy soil slightly acid with humus, rich in minerals and containing some nutrients. Found in mountain regions from south Wales and the Midlands northwards.

Additional information
This plant used to be very popular as an ornamental plant and as a wild flower for posies.

Mignonette family *Resedaceae*
Flowering time: June–Aug.
Height of growth: 30–75 cm
Dicotyledonous; Biennial – Perennial

Identification marks
Stem often branched, ascending to erect. Upper leaves pinnate or bipinnate. Dense racemes. Flowers pale yellow; usually 4 large deeply cleft petals, plus 2 very small ones.

Habitat and distribution
Rubble heaps, paths, railway embankments, in fields. Calcicolous. Needs disturbed, often stony ground rich in nutrients. Throughout most parts.

Additional information
Similar to Weld (*R. luteola*), common in similar habitats. Old dye plant. Always has only 4 petals, all leaves undivided.

Goldenrod
Solidago virgaurea

Canadian Goldenrod
Solidago canadensis

Daisy family *Asteraceae (Compositae)*
Flowering time: July–Sept.
Height of growth: 5–75 cm
Dicotyledonous; Perennial

Identification marks
Many small capitula, 7–10 mm long, in an erect raceme or panicle. On the outside 5–12 narrow ray florets, on the inside disc florets. Stem erect, branches rod-like. Lower leaves obovate, toothed, upper ones ovoid-lanceolate.

Habitat and distribution
Open woodland, glades, heaths, poor turf, dunes, cliffs. Likes sunny, preferably calcareous soils which are low in nutrients but not too dry. Common; found up to 1,200 m.

Daisy family *Asteraceae (Compositae)*
Flowering time: Aug.–Oct.
Height of growth: 50–250 cm
Dicotyledonous; Perennial

Identification marks
Hundreds of small capitula, about 5 mm long, in curved, 1-sided partial inflorescences, together forming a projecting panicle. Ray florets are hardly longer than disc florets. Stem erect, usually not branched (apart from the projecting flower-bearing branches) hairy. Leaves lanceolate, often somewhat serrate.

Habitat and distribution
Widely cultivated and a frequent escape, in many areas. On sunny but damp loamy soils rich in nutrients.

Additional information
Very similar: Early Goldenrod (*S. gigantea = serotina*), equally tall, base of stem hairless, often tinged with red; ray florets clearly longer than disc florets. Somewhat rarer as an escape. Both introduced from N. America and established here as ornamentals.

Irish Fleabane
Inula salicina

Ploughman's-spikenard
Inula conyza

Daisy family *Asteraceae (Compositae)*
Flowering time: July–Aug.
Height of growth: 30–50 cm
Dicotyledonous; Perennial

Identification marks
Stem erect, with 1–5 capitula 2.5–3 cm wide. On the outside long ray florets, inside disc florets. Fruit with a pappus of hairs. Leaves alternate, margin entire, usually hairless, or hairy on veins of underside only.

Habitat and distribution
Semi-dry turf, thickets, balks, damp meadows. On calcareous sometimes damp soils. This species only found outside continental Europe on the limestone shoreline of Lough Derg in Galway and Tipperary.

Additional information
Similar: Fleabane (*Pulicaria dysenterica*), capitula smaller, leaves hairy. Yellow Ox-Eye (p.162): capitula wider, leaves hairy. Further *Inula* species whose leaves are all conspicuously hairy at least on the undersides.

Daisy family *Asteraceae (Compositae)*
Flowering time: July–Sept.
Height of growth: 20–130 cm
Dicotyledonous; Biennial – Perennial

Identification marks
Numerous capitula approx. 1 cm wide form a terminal corymb. Only disc florets. Bracts have spreading tips. Stem usually erect, reddish-brown. Leaves ovoid-oblong, finely serrated, undersides covered with felt-like downy hairs.

Habitat and distribution
Open woods, glades, thickets, high moors, walls, and cliffs. Calcicolous. Found on sunny and rather dry stony ground where the soil is not too poor. Local, occasionally common in England and Wales.

Additional information
Similar: *A. graveolens* (= *Cupularia graveolens*), annual, unpleasant smell, glandular, sticky, narrow leaves. Very rare in mild locations in salty areas.

Yellow Ox-eye
Buphthalmum salicifolium

Jerusalem Artichoke
Helianthus tuberosus

Daisy family *Asteraceae (Compositae)*
Flowering time: July–Sept.
Height of growth: 20–60 cm
Dicotyledonous; Perennial

Identification marks
Capitula 3–6 cm wide, florets, inside disc florets. Pappus a scarious rim with minute teeth. Stem erect, little branching. Leaves alternate, lanceolate, covered with soft hairs; upper leaves sessile.

Habitat and distribution
Not British. Open woodland, dry thickets, forest margins, heaths, poor turf. On calcareous and often stony ground.

Additional information
There are several similar species from the Fleabane genus (*Pulicaria*) and that of *Inula* (see p.161). However only the Ox-eye has paleae (large scales) in the capitula.

Daisy family *Asteraceae (Compositae)*
Flowering time: Sept.–Nov.
Height of growth: 1–2.5 m
Dicotyledonous; Perennial

Identification marks
Capitula 3–8 cm wide, erect, terminal. On the outside ray florets, inside disc florets. Stem erect, little branching. Lower leaves opposite, ovoid to heart-shaped, upper ones alternate, narrow-ovoid; all petiolate, toothed margin, leaf blade covered with rough hairs.

Habitat and distribution
Widely grown as a vegetable for its edible tubers, often escaping. Grows on sandy or gravelly loamy soil rich in nutrients. Native of North America.

Additional information
Occasionally other related species (ornamental plants) are found growing wild; definition is very difficult.

Trifid Bur-marigold
Bidens tripartita

Pineappleweed
Chamomilla suaveolens

Daisy family *Asteraceae (Compositae)*
Flowering time: July–Sept.
Height of growth: 15–60 cm
Dicotyledonous; Annual

Identification marks
Capitula solitary, 1.5–2.5 cm wide and equally long. On the outside occasionally ray florets, inside brownish-yellow disc florets. Fruit usually has 2 barbed bristles. Stem erect, often tinged brownish-red. Leaves opposite, 3–5 lobed.
Habitat and distribution
Ditches, river banks, wet meadows and fields. Found on wet sandy or muddy ground rich in nutrients.
Additional information
Similar: Greater Bur-marigold (*B. radiata*), not British, capitula larger, leaves light green; rare. *B. trondosa* (= *melanocarpa*), leaves pinnate with 3–5 leaflets and long petioles. An American introduction which occasionally becomes established.

Daisy family *Asteraceae (Compositae)*
Flowering time: June–July
Height of growth: 5–40 cm
Dicotyledonous; Annual

Identification marks
Capitula semi-spherical, greenish-yellow. Usually only disc florets, very rarely stunted white ray florets. Stem prostate to erect, many branches, glabrous, very leafy. Leaves double pinnate with narrow lobes. The entire plant has an aromatic smell.
Habitat and distribution
Paths, fallow land, rubble heaps, railway embankments, walls, sports fields, more rarely in cornfields. Likes rather damp soil rich in nutrients which is not overgrown but is compact as a result of constant passage. Grows well near residential areas. Very common.
Additional information
Synonyms: *Matricaria matricarioides, M. suaveolens*. Introduced into Europe only in 1850.

Tansy
Tanacetum vulgare

Mugwort
Artemisia vulgaris

Daisy family *Asteraceae (Compositae)*
Flowering time: July–Sept.
Height of growth: 60–120 cm
Dicotyledonous; Perennial

Identification marks
Capitula in corymbs, semi-spherical and flattened, approx. 1 cm wide. All flowers composed only of disc florets, the marginal florets occasionally have very short ligules. Pappus of the fruit a short, papery cup. Stem erect, angled. Leaves alternate, pinnate, with serrate lobes. Slightly harsh aromatic smell.
Habitat and distribution
Pathways, rubble heaps, embankments, waste places. Likes loamy soil rich in nutrients in warm summer locations but with adequate rainfall. Common.
Additional information
Synonyms: *Chrysanthemum tanacetum, C. vulgare*

Daisy family *Asteraceae (Compositae)*
Flowering time: July–Sept.
Height of growth: 60–120 cm
Dicotyledonous; Perennial

Identification marks
Capitula narrow, approx. 5 mm long, numerous, in leafy racemose panicles. Disc florets only, virtually entirely enclosed by the somewhat felted involucral bracts. Stem erect, often tinged with dark reddish-violet. Leaves pinnate, undersides covered with white downy hairs. Aromatic smell.
Habitat and distribution
Pathways, rubble heaps, embankments, roadsides, waste places. On various different nitrogenous soils. Very common.
Additional information
Old medicinal and spice plant.

Colt's-foot
Tussilago farfara

Arnica
Arnica montana

Daisy family *Asteraceae (Compositae)*
Flowering time: March–April
Height of growth: 10–30 cm
Dicotyledonous; Perennial

Identification marks
Solitary capitula, terminal on stem with scale-like leaves. On the outside several rows of ray florets, on the inside disc florets. Fruit with a pappus of long white hairs. After flowering, stem is longer, and drooping at the top. Basal leaves appear after flowering time: long-stalked, round to heart-shaped, toothed, undersides covered with white felt; teeth slightly black.

Habitat and distribution
Arable land, open ground, paths, embankments, walls, waste places. Calcicolous, requires moisture. Very common.

Additional information
Old medicinal plant. The similar leaves of the Butterbur (*Petasites*, pp.88 and 243) do not have blackish teeth.

Daisy family *Asteraceae (Compositae)*
Flowering time: une–July
Height of growth: 20–50 cm
Dicotyledonous; Perennial

Identification marks
Stem erect, usually unbranched, with 1–2 pairs of opposite leaves, downy. Rosette of leathery, ovoid basal leaves with margin almost entire. All capitula 6–8 cm wide; on the outside long ray florets, on the inside disc florets. Fruit with a pappus of hairs. Slightly harsh aromatic smell.

Habitat and distribution
Not British. Poor meadowland, Alpine mats, moorland, heaths. Likes acid loamy soil containing sand and humus; also peaty soil. Calcifugous. Rare. Main area of distribution is the mountains.

Additional information
Old medicinal plant. Can be distinguished from other yellow *Compositae* by its basal rosette and its opposite leaves.

Groundsel
Senecio vulgaris

Sticky Groundsel
Senecio viscosus

Daisy family *Asteraceae (Compositae)*
Flowering time: Jan.–Dec.
Height of growth: 10–45 cm
Dicotyledonous; Perennial

Identification marks
Capitula arranged in long corymbose clusters. Only disc florets; barely longer than the involucral bracts which have black specks. Fruit with a pappus of hairs. Stem ascending-erect, branched. Leaves alternate, pinnately lobed, often with a cobweb-like covering of hairs.

Habitat and distribution
Gardens, fields, paths, rubble heaps, clearings. Found on various soils, all of which are, however, nitrogenous and slightly damp. Very common throughout.

Additional information
After flowering the small hairy fruits make the capitula look like a white-haired old man (Latin: *Senecio – senex = old man*).

Daisy family *Asteraceae (Compositae)*
Flowering time: July–Sept.
Height of growth: 10–60 cm
Dicotyledonous; Annual

Identification marks
Capitula arranged in long-ovoid corymb. On the outside narrow ray florets which are usually rolled back, on the inside disc florets. The inflorescence at least is sticky-glandular. Fruit with a pappus of hairs. Stem usually erect. Leaves pinnately lobed with lanceolate coarsely toothed lobes.

Habitat and distribution
Rubble heaps, clearings, dunes, railway embankments, pathways. Likes very stony open ground preferably low in lime and not too damp. Scattered, sometimes common throughout lowland areas.

Additional information
Very similar: Heath Groundsel, *S. sylvaticus*, not sticky. Found in similar habitats. Scattered.

Senecio vernalis

Senecio helenitis

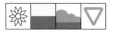

Daisy family *Asteraceae (Compositae)*
Flowering time: May–Nov.
Height of growth: 15–50 cm
Dicotyledonous; Annual

Identification marks
Capitula almost campanulate, in a more or less corymbose arrangement. On the inside disc florets, on the outside 13 projecting ray florets. Fruit with a pappus of hairs. Stem erect, little branching. Leaves pinnately lobed, the upper amplexicaul, margin completely serrate.

Habitat and distribution
Not British. Fields, paths, rubble heaps, forest margins, railway embankments. Usually found on somewhat dry sandy soil rich in nitrogen but rather low in lime and warm in the summer. Tolerates salt. Rare and usually only temporary but repeatedly introduced (from the East) from approximately 1850 with a steady trend towards a spread westwards in northern Europe.

Butterwort family *Asteraceae (Compositae)*
Flowering time: May–June
Height of growth: 50–100 cm
Dicotyledonous; Perennial

Identification marks
Capitula campanulate, in a loose corymb. On the outside 13 projecting ray florets, on the inside disc florets. Fruit with a pappus of hairs. Stem stiff and erect. Leaves ovoid-spatulate, usually with margin curled up, woolly and cobweb-like; stem leaves very narrow, lower ones amplexicaul.

Habitat and distribution
Not British. Marshy meadowland, open woodland. Likes peaty soils which are low in nutrients and free of lime and occasionally wet. Rare, found only in the south and west of northern Europe.

Additional information
Synonym: *S. spathulifolius*. Many subspecies.

Hoary Ragwort
Senecio erucifolius

Wood Ragwort
Senecio nemorensis

Daisy family *Asteraceae (Compositae)*
Flowering time: July–Aug.
Height of growth: 30–120 cm
Dicotyledonous; Biennial – Perennial

Identification marks
Capitula in terminal and axillary corymbs. On the outside (12–14) projecting ray florets, on the inside disc florets. Fruit with a pappus of hairs. Stem grows erect from short horizontal rootstock. Often reddish-brown, angular. Leaves pinnately lobed, covered with woolly web on the undersides.
Habitat and distribution
Balks, pathways, thickets, shingle. Likes nitrogenous loamy soil which is stony and preferably calcareous. Locally common in England and Wales, rarer in a few parts of southern Scotland.
A: Very similar: Common Ragwort (*S. jacobaea*). Rootstock goes straight into the ground: capitula 15–25 mm wide (instead of 10–15 mm).

Daisy family *Asteraceae (Compositae)*
Flowering time: July–Aug.
Height of growth: 50–150 cm
Dicotyledonous; Perennial

Identification marks
Capitula in umbel-like corymbs. On the outside 5 or 7 ray florets, on inside disc florets. Fruit with pappus of hairs. Stem erect, leaves alternate. Leaves ovoid-lanceolate, serrate, at most with short petioles.
Habitat and distribution
Not British. Woodland, undergrowth along streams, glades, brush areas in clearings and Alpine regions. Needs well-moistened soil, rich in nutrients and humus. Scattered; rare towards northern Europe.
Additional information
Ssp. *nemorensis*, type found in moderately mountainous regions of central and eastern Europe, stem green, hairy; leaf margin ciliate. Ssp. *fuchsii*, from the lowlands up to 2,000 m; stem red, glabrous, leaves narrow, at least 4 times longer than wide.

Field Marigold
Calendula arvensis

Carline Thistle
Carlina vulgaris

Daisy family *Asteraceae (Compositae)*
Flowering time: May–Oct.
Height of growth: 10–20 cm
Dicotyledonous; Annual

Identification marks
Capitula 1–2 cm wide on long peduncles. On the outside ray florets, on the inside disc florets. Fruit is bent over in the shape of a hook or ring and the back of it is covered with spines. Leaves spatulate to lanceolate, amplexicaul, with covering of downy hairs.

Habitat and distribution
Prefers dry soil rich in nutrients and minerals. A frequent casual but not becoming established.

Additional information
Similar: Pot Marigold (*C. officinalis*), old medicinal and ornamental plant, often escaping. Capitula 2–5 cm across, orange, sometimes yellow.

Daisy family *Asteraceae (Compositae)*
Flowering time: July–Oct.
Height of growth: 15–60 cm
Dicotyledonous; Biennial

Identification marks
Leaves long, margin lobed with prickly thorns. Stem erect, usualy branched. Capitula 2–4 cm wide, in corymbs, occasionally solitary. Bracts thorny, the inner row ray-like and straw-coloured. Only disc florets present. Fruit with a pappus of hairs.

Habitat and distribution
Dry grassland. Calcicolous. Likes somewhat dry soil rich in minerals but low in nutrients. Locally common, rarer in the far north and high regions.

Cabbage Thistle
Cirsium oleraceum

Daisy family *Asteraceae (Compositae)*
Flowering time: July–Sept.
Height of growth: 50–120 cm
Dicotyledonous; Perennial

Identification marks
Leaves glabrous, serrate with soft prickles, light green; upper ones often undivided, lower ones pinnately divided. Capitula erect, 2–4 cm long growing terminally in small groups and surrounded by pale bracts. Only disc florets present. Fruit with a pappus of hairs.

Habitat and distribution
Damp meadows, flat moorland, river banks, ditches, wet woods. Calcicolous. Indicator of wetness. Likes loamy soil rich in nutrients. An introduction from central Europe, established in a few places.

Cirsium spinosissimum

Daisy family *Asteraceae (Compositae)*
Flowering time: July–Aug.
Height of growth: 20–50 cm
Dicotyledonous; Perennial

Identification marks
Stem erect, dense foliage. Leaves yellowish-green, pinnately lobed, margin serrate and thorny. Capitula 2–3 cm long; these are usually terminal in small groups and surrounded by pale yellow thorny involucral bracts. Fruit with a pappus of hairs.

Habitat and distribution
Not British. Only found in Alpine regions of Europe (France to Yugoslavia). On damp mats and stony turf, grazing land and brush areas near river banks. Likes nitrogen. Found on damp loamy soil. Scattered.

Additional information
Various habitat-related types: in the shade grows up to over I m high; in higher locations single capitulum or very short stem (5 cm) Growth is promoted by grazing.

Rough Hawkbit
Leontodon hispidus

Hawkweed Oxtongue
Picris hieracioides

Daisy family *Cichoriaceae (Compositae)*
Flowering time: June–Sept.
Height of growth: 10–60 cm
Dicotyledonous; Perennial

Identification marks
Capitula terminal, 2.5–4 cm wide. Ray florets only. Stem slender, pithy, usually unbranched; scale-like leaves, 1–2 at most. Basal leaves pinnately lobed or entire. Plant contains a milky juice (latex).

Habitat and distribution
Meadows, balks, pastures. Found on soil rich in nutrients. Found throughout Britain, sometimes common.

Additional information
Very variable, also in the hair covering. Similar: Autumn Hawkbit (*L. autumnalis*); usually branched, leaves mostly glabrous, toothed to pinnately lobed; meadows. Abundant. Cat's-ear (*Hypochoeris radicata*): stem bluish-green. Leaves coarsely serrate, scattered bristles; grassland. Common.

Daisy family *Cichoriaceae (Compositae)*
Flowering time: July–Sept.
Height of growth: 15–90 cm
Dicotyledonous; Biennial – Perennial

Identification marks
Leaves and at least lower part of stem covered with stiff bristles. Plant contains milky juice (latex). Stem usually richly branched and squarrose. Leaves long, coarsely toothed. Capitula in corymbs, 2–3.5 cm wide. Outer involucral bracts spreading. Ray florets only. Fruit with a pappus of hairs.

Habitat and distribution
Grassland, wayside and grassy areas. Grows on calcareous loamy ground which is not too dry but rich in nutrients and preferably stony. Local, sometimes common in the lowlands of England, Wales and southern Scotland.

Goat's beard
Tragopogon pratensis

Dandelion
Taraxacum officinale

Daisy family *Cichoriaceae (Compositae)*
Flowering time: June–July
Height of growth: 30–70 cm
Dicotyledonous; Annual – Biennial

Identification marks
Capitula 4–6 cm wide. Only ray florets. Fruit with a pappus of hairs. Only 1 row of bracts. Stem erect, little branching, slightly swollen under the capitula. Leaves opposite, margin entire, sheathing stem. Abundant latex.

Habitat and distribution
Meadows; rarely along paths. Likes well-moistened loamy soil containing nutrients and minerals. Locally common more or less throughout.

Additional information
Ssp. *pratensis*, rays pale yellow, as long as the involucral bracts. Common in Europe, less so in Britain. Ssp. *minor*, rays bright yellow, only as long as the involucral bracts. The most common in Britain. Ssp. *orientalis*, rays golden yellow, longer than the involucral bracts. Casual in Britain.

Daisy family *Cichoriaceae (Compositae)*
Flowering time: April–June
Height of growth: 10–60 cm
Dicotyledonous; Perennial

Identification marks
Capitula 2–5 cm wide, solitary on leafless, wide, hollow peduncles. Only ray florets. Two rows of involucral bracts, the outer ones often recurved. Fruit with a pappus of hairs. Basal leaves in a rosette, long, jaggedly serrate. Plant contains milky juice (latex).

Habitat and distribution
Everywhere in habitats which are not too wet and shady. Very common; up to around 1,900 m.

Additional information
Collective species or aggregate; numerous micro-species have been recognized.

Perennial Sow-thistle
Sonchus arvensis

Smooth Sow-thistle
Sonchus oleraceus

Daisy family *Cichoriaceae (Compositae)*
Flowering time: July–Oct.
Height of growth: 60–150 cm
Dicotyledonous; Perennial

Identification marks
Capitula up to 5 cm wide, in a loose corymb. Only ray florets. Inflorescence branches and involucral bracts covered with dense layer of yellow glandular hairs. Fruit with a pappus of hairs. Stem only branched at the top, hollow. Leaves jaggedly serrate, the upper ones less so, all spiny serrate. Plant contains abundant milky juice (latex).

Habitat and distribution
Fields, gardens, waysides, rubble heaps, salty marshes. Found on nitrogenous soil. Common throughout. Tolerates salt, needs warmth.

Additional information
Similar: Marsh Thistle (*S. palustris*), glands black, capitula up to 4 cm across.
River banks, marshes, etc. Rare, south-east England only.

Daisy family *Cichoriaceae (Compositae)*
Flowering time: May–July
Height of growth: June–Aug.
Dicotyledonous; Annual

Identification marks
Plant is usually glabrous, with abundant milky juice (latex). Capitula 2–2.5 cm wide, in cymose umbels. Only ray florets. Fruit with a pappus of hairs. Stem branched, hollow. Leaves variable, those of the stem usually pinnately divided or serrate, not spiny, with projecting pointed auricles clasping at the stem.

Habitat and distribution
A weed of cultivated and waste places. Found on ground which is rich in nitrogen but not too dry. Common throughout.

Additional information
The flower colour ranges from rich yellow to whitish-yellow. Very similar: Prickly Sow-thistle (*S. asper*), leaf auricles have rounded apices, pressed close to the stem. Common in similar habitats.

Wall Lettuce
Mycelis muralis

Nipplewort
Lapsana communis

Daisy family *Cichoriaceae (Compositae)*
Flowering time: July–Sept.
Height of growth: 30–100 cm
Dicotyledonous; Perennial

Identification marks
Capitula in panicles, barely 1 cm wide. Only 5 ray florets per capitulum, pale yellow. Fruit with a pappus of hairs. Stem erect, branched towards the top. Leaves frequently tinged with dirty red colour, the lower ones petiolate, the upper ones sessile with large terminal lobe. Plant glabrous and containing abundant milky juice (latex).

Habitat and distribution
On shady rocks and walls, occasionally in deciduous woodland. Likes loose loamy soil rich in nutrients with layer of mull. Can also be stony. Scattered in most areas, absent from some parts, notably the Highlands of Scotland.

Additional information
Synonyms: *Prenanthes muralis, Lactuca muralis.*

Daisy family *Cichoriaceae (Compositae)*
Flowering time: July–Sept.
Height of growth: 20–90 cm
Dicotyledonous; Perennial

Identification marks
Capitula arranged in panicles, approx. 1.5–2 cm wide. Only 8–15 light yellow ray florets per capitulum. Fruit without a pappus. Stem erect, branched towards the top. Lower leaves pinnately lobed, with large terminal lobe, upper ones ovoid, almost all petiolate. Plant contains milky juice (latex) and is usually hairy.

Habitat and distribution
Ploughed fields, gardens, rubble heaps, also in thickets, along wood margins, in glades and in light open woodland. Likes somewhat damp soil rich in nutrients. Common throughout.

Additional information
The plant varies a great deal in its hair covering, its leaf shape and its size.

Rough Hawk's-beard
Crepis biennis

Smooth Hawk's–beard
Crepis capillaris

Daisy family *Cichoriaceae (Compositae)*
Flowering time: June–July
Height of growth: 30–120 cm
Dicotyledonous; Biennial

Daisy family *Cichoriaceae (Compositae)*
Flowering time: June–Sept.
Height of growth: 20–90 cm
Dicotyledonous; Annual – Biennial

Identification marks
Capitula 2–3.5 cm wide, in corymbs. Only ray florets. Outer involucral bracts spreading. Fruit with a pappus of hairs. Stem erect, grooved, branched at the top. Leaves pinnately divided, upper ones undivided. Plant contains milky juice (latex).

Habitat and distribution
Meadows, path balks, rarely found on fields. Likes loamy soil rich in nutrients. Local, somewhat rarer towards the north.

Additional information
Similar: Northern Hawk's-beard (*C. mollis*), outer involucral bracts pressed inwards, leaf margin is entire to serrate. Mountain meadows and pastures. Local, rare in the lowlands.

Identification marks
Capitula 1–1.5 cm wide, arranged in loose corymbs. Only ray florets. Fruit with a pappus of hairs. Stem branched above. Leaves jaggedly serrate to pinnately divided, upper ones narrow, all sagittate at the base, amplexicaul. Plant is usually glabrous and contains milky juice (latex).

Habitat and distribution
Meadows, pastures, park lawns, pathway balks. Found on soil which is not too damp and is relatively low in nutrients, preferably also low in lime. Common throughout.

Additional information
This is the most common of a number of Hawk's–beard species which are small in growth and very difficult to distinguish from one another. Synonym: *C. virens*.

Prickly Lettuce
Lactuca serriola

Mouse-ear Hawkweed
Hieracium pilosella

Daisy family *Cichoriaceae (Compositae)*
Flowering time: July–Sept.
Height of growth: 30–150 cm
Dicotyledonous; Annual – Biennial

Identification marks
Capitula approx. 1 cm wide, in panicles. Only ray florets. Fruit with a pappus of hairs. Stem whitish, often tinged with reddish-violet. Leaves undivided to pinnately divided, sagittate at the base, toothed, prickly beneath; leaves on stem set perpendicularly and arranged in a north-south direction; central rib on leaf underside is prickly. Plant glabrous; contains abundant milky juice (latex).

Habitat and distribution
Paths, railway embankments, rubble heaps, wasteland. Found on warm soil rich in nitrogen. South and central England and Wales.

Additional information
Synonym: *L. scariola*.

Daisy family *Cichoriaceae (Compositae)*
Flowering time: May–Aug.
Height of growth: 5–30 cm
Dicotyledonous; Perennial

Identification marks
Capitula solitary on leafless stems, 2–3 cm wide. Only ray florets. Fruit with a pappus of hairs. Leaves basal, ovoid, undersides covered with felt-like hairs, uppersides have individual long bristles. Plant contains milky juice (latex). Has long runners above the ground.

Habitat and distribution
Poor turf, dry turf, balks, heaths, open woodland. Found on various soils. Avoids wet and shady localities. Found throughout Britain.

Hieracium murorum group

Daisy family *Cichoriaceae (Compositae)*
Flowering time: May–Oct.
Height of growth: 10–40 cm
Dicotyledonous; Perennial

Identification marks
Capitula 2–3 cm wide, corymbosely arranged in small numbers. Only ray florets. Fruit with a pappus of hairs. Stem erect, with no leaves (or up to 2). Remaining leaves are basal, more or less ovoid, usually somewhat toothed, petiolate. No runners. Plant contains milky juice (latex), usually (glandular) hairy. Hairs are black.

Habitat and distribution
All types of woodland, less common in the shadow of rocks and walls. Found on soil which is low in lime, not too dry but rich in humus. Common in central and southern England and Wales, rare in southern Scotland.

Additional information
Synonym: *H. sylvaticum*. Very many closely related species.

Umbellate Hawkweed group
Hieracium umbellatum

Daisy family *Cichoriaceae (Compositae)*
Flowering time: July–Sept.
Height of growth: 30–80 cm
Dicotyledonous; Perennial

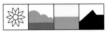

Identification marks
Capitula 1–3 cm wide, in an umbellate panicle. Only ray florets. Tips of the involucral bracts are recurved. Fruit has ciliate margin. Many narrow stem leaves. Plant contains milky juice (latex)(but often only very little).

Habitat and distribution
Wood margins, thickets, heaths, poor turf, dunes. Found on loamy soil low in lime and frequently sandy. Found throughout but chiefly in the lowland areas.

Additional information
Similar: Species of the *H. sabauda* group: leaves ovoid, involucral bracts pressed inwards.

Lady's-slipper
Cypripedium calceolus

Pale-flowered Orchid
Orchis pallens

Orchid family *Orchidaceae*
Flowering time: May–June
Height of growth: 15–45 cm
Monocotyledonous; Perennial

Identification marks
Stem erect, with 1–2 (rarely as many as 4) flowers. Large yellow labellum, 3–4 cm long inflated and slipper-like; plus 4 lanceolate, purple-brownish outer perianth segments, 4–6 cm long. 3–4 leaves, large, elliptical, with sheathing bases. Leaf margin ciliate.

Habitat and distribution
Deciduous woodland. Found on loamy soil rich in lime but not too dry. Very rare, has been wiped out in many areas by collectors. Nowadays found only in one place in Yorkshire.

Orchid family *Orchidaceae*
Flowering time: April–May
Height of growth: 15–40 cm
Monocotyledonous; Perennial

Identification marks
Short, somewhat dense spike. Outer petals spreading, flower spur at most as long as the ovary, projecting horizontally or curved upwards. Bracts membraneous, pale yellow with a single vein. Leaves broadly ovoid, not spotted.

Habitat and distribution
Not British. Deciduous woodland, sunny mixed woodland. Found on somewhat moist calcareous soil containing humus. Likes warmth. Rare. Found in Germany, Austria, Alps and southern Europe; in the Alps up to approx. 1,200 m.

Additional information
Do not confuse with the following entry, Elder-flowered Orchid, even though the flowers of the Pale Orchid have similar unpleasantly heavy aroma of elder.

Elder-flowered Orchid
Dactylorhiza sambucina

Bird's-nest Orchid
Neottia nidus-avis

Orchid family *Orchidaceae*
Flowering time: April–June
Height of growth: 15–25 cm
Monocotyledonous; Perennial

Identification marks
Short compressed spike. Outer petals projecting, flower spur longer than ovary. Flower bracts deciduous (not membraneous). Leaves long-ovoid, unspotted.

Habitat and distribution
Not British. On sunny balks, poor turf, in open woodland and thickets. Prefers moderately dry, stony, loamy soil low in lime. Very rare.

Additional information
The spike can be any shade from red to yellow but the yellow plants are more common. Similar: many types of *Orchis* and *Dactylorhiza (= Dactylorchis)* with yellow or red flowers.

Orchid family *Orchidaceae*
Flowering time: June–July
Height of growth: 20–45 cm
Monocotyledonous; Perennial

Identification marks
Entire plant yellowish-brown. Tough stem with narrow-ovoid leaves. Raceme with many flowers, cylindrical. Flowers do not have a spur. Labellum bi-lobed, upper perianth-segments coming together to form a hood.

Habitat and distribution
In deciduous and mixed woodland especially in beechwood. Likes loamy soils rich in lime and nutrients and absorbs organic material in the mull with the help of fungi. Throughout Britain but easily overlooked.

Additional information
The rootstock under the ground bears many tightly interwoven roots. Their nest-like appearance gave the plant its name.

Birthwort
Aristolochia clematitis

Wolf's-bane
Aconitum vulparia

Birthwort family *Aristolochiaceae*
Flowering time: June–Sept.
Height of growth: 20–80 cm
Dicotyledonous; Perennial

Identification marks
Stem erect or slightly twining, glabrous. Leaves alternate, the blades about twice as long as the petioles, round-ovoid, with heart-shaped base. Flowers in leaf axils, inflated and globular at the base then becoming a long straight tube terminating in a tongue-shaped lobe; the inside of the tube is covered with hairs pointing inwards (fly-trap).

Habitat and distribution
Waysides, walls, thickets. Likes the warmth. Calcicolous. Local in eastern and central England and a few places elsewhere.

Additional information
Old medicinal and poisonous plant from the Mediterranean region; in Britain it has grown wild from old gardens and has succeeded in establishing itself.

Buttercup family *Ranunculaceae*
Flowering time: April–May
Height of growth: 50–150 cm
Dicotyledonous; Perennial

Identification marks
Flowers in simple or branched raceme, the topmost petal forming a tall cylindrical helmet 1.5–2 cm long, which is closed off below by 4 smaller oval-shaped petals. Stem erect, with sparse hairs below, more dense further up. Leaves alternate, palmately lobed with 5–7 segments, the lower ones having long petioles. Leaf segments themselves 3-lobed and coarsely serrate.

Habitat and distribution
Not British. Moist deciduous woodland, canyon forests, brakes and lowland forests. Likes damp humus soil rich in nutrients. Rare. Main areas of distribution are the higher mountainous regions of C. and S. Europe.

Additional information
Very many different varieties. Synonym: *A. lycoctonum*.

Yellow Corydalis
Corydalis lutea

Winged Broom
Chamaespartium sagittale

Poppy family *Papaveraceae*
Flowering time: May–Aug.
Height of growth: 10–30 cm
Dicotyledonous; Perennial

Identification marks
Stem ascending to erect, richly branched, very leafy. Leaves petiolate, pinnate, the leaflets divided into 3–5 wedge-shaped segments which are usually toothed or lobed at the apex. Entire plant glabrous. Dense unilateral racemes. Flowers up to 2 cm long, with a short spur, golden yellow, rich yellow at the tip.

Habitat and distribution
On walls and rocks. Ornamental plant from the Mediterranean region which has grown wild and established itself in a number of places. Likes moist calcareous cracks containing some nutrients; does not like full exposure to the sun. Scattered throughout.

Pea family *Fabaceae(Leguminosae)*
Flowering time: May–June
Height of growth: 10–30 cm
Dicotyledonous

Identification marks
Branches ascending to erect, with broad wings, sparse foliage. Leaves alternate, sessile, ovoid-lanceolate, margin entire, easily deciduous. Entire plant covered with hairs, but rapidly becoming glabrous, then tough and leathery and woody at the base. Pea flowers 10–15 mm long. Flowers in dense terminal racemes.

Habitat and distribution
Not British. Semi-dry turf, poor meadowland, path balks, heaths, open dry woodland, sometimes also on rocks. Somewhat calcifugous. Indicator of surface acidity. Common in central and southern Europe, but rarer in the north.

Additional information
Has been classified under various different names: *Genista sagitalis, Cytisus sagitalis*.

Wild Liquorice
Astragalus glycyphyllos

Ribbed Melilot
Melilotus officinalis

Pea family *Fabaceae (Leguminosae)*
Flowering time: July–Aug.
Height of growth: 60–100 cm
Dicotyledonous; Perennial

Identification marks
Stem creeping to ascending, only slightly hairy. Leaves alternate, imparipinnate with 10–20 long-oval leaflets. Flowers pale, (greenish-) yellow to ivory-coloured. Racemes in leaf axils on short peduncles. Fruit is a somewhat puffy pod, curved almost to the point of being spiral.
Habitat and distribution
Open dry grassy areas and dry thickets. Likes a loamy soil which is both nitrogenous and calcareous. Scattered throughout Britain.
Additional information
Medicinal plant rich in sugars.

Pea family *Fabaceae (Leguminosae)*
Flowering time: July–Sept.
Height of growth: 60–120 cm
Dicotyledonous; Biennial

Identification marks
Stem ascending to erect, heavily branched. Leaves alternate, trifoliate; leaflets long-ovoid, serrate. Dense somewhat unilateral racemes with long peduncles in the axils of the upper leaves. 30–60 flowers, only 5–6 mm long, the wings and standard longer than the keel. Ovary and fruit glabrous.
Habitat and distribution
Pathways, railway ballast, fields, wasteland, quarries, river banks, rubble heaps. Nitrogen indicator. Found in somewhat dry sunny places. Southern England only.
Additional information
Very similar: Tall Melilot (*Melilotus altissima*), fruit hairy (wings, standard and keel more or less the same length). Scattered throughout most areas.

Sickle Medick
Medicago sativa ssp. *falcata*

Black Medick
Medicago lupulina

Pea family *Fabaceae (Leguminosae)*
Flowering time: June–July
Height of growth: 30–60 cm
Dicotyledonous; Perennial

Identification marks
Flowers approx. 1 cm long, 6–20 in pedunculate racemes coming out of the axils of the alternate leaves. Stem escending-erect. Leaves made up of 3 leaflets which are long and emarginate at the apex. Pods are usually sickle-shaped.

Habitat and distribution
Dry turn, poor meadowland. Prefers rather dry calcareous soil. Only in the Breckland (East Anglia).

Additional information
Sometimes separated from Lucerne (p.332) as a species in its own right.

Pea family *Fabaceae (Leguminosae)*
Flowering time: April–Aug.
Height of growth: 5–50 cm
Dicotyledonous; Annual – Perennial

Identification marks
Stem prostrate to erect, angular. Leaves trifoliate, leaflets have hairy undersides. Flowers 2–3 mm long, initially in compact, globular racemes of 10–50 flowers, approx. 5 mm across. Petals fall off after flowering. Pods are kidney-shaped.

Habitat and distribution
Dry meadowland, poor turf, path balks, railway ballast. Calcicolous. Likes warm soil rich in nitrogen. Common in most areas.

Additional information
Very similar: yellow flowering species of Trefoil (*Trifolium*, see following pages); they usually have less hair and the petals remain brown and dried up in the calyx.

Kidney Vetch
Anthyllis vulneraria

Large Hop Trefoil
Trifolium aureum

Pea family *Fabaceae (Leguminosae)*
Flowering time: June–Sept.
Height of growth: 10–60 cm
Dicotyledonous; Perennial

Identification marks
Stem prostrate to erect. Basal leaves sometimes consisting of the terminal leaflet only. Stem leaves imparipinnate; leaflets long, terminal leaflets much larger than the lateral ones which are sometimes absent. Flowers in dense cymes with pinnately divided bracts.

Habitat and distribution
Dry poor turf, balks. Somewhat calcicolous and commoner near the sea. Throughout Britain.

Additional information
Many subspecies. Ssp. *vulneraria*, hairy, flowers shiny yellow. Ssp. *maritima*, flowers orange-yellow, grows on sand along the North Sea and Baltic coasts. Ssp. *alpestris*, mountain form; flowers large, calyx covered with greyish hairs, flowers golden yellow. Alpine grassland.

Pea family *Fabaceae (Leguminosae)*
Flowering time: July–Aug.
Height of growth: 10–50 cm
Dicotyledonous; Annual

Identification marks
Stem usually erect, richly branched right from the bottom. Leaves alternate, trifoliate; leaflets ovoid, all with short, equal petioles. 20–50 flowers in a somewhat long, 7–10 mm wide head. After flowering the petals remain on the flower, dry and yellowish-brown (used for aerial distribution of seeds).

Habitat and distribution
Dry turf, poor meadowland, wayside verges, balks. Somewhat calcifugous. Indicator of slight surface acidity. Scattered throughout and somewhat rare.

Additional information
Synonyms: *T. strepens*, *T. agrarium*.

Hop Trefoil
Trifolium campestre

Lesser Trefoil
Trifolium dubium

Pea family *Fabaceae (Leguminosae)*
Flowering time: June–Sept.
Height of growth: 5–30 cm
Dicotyledonous; Annual

Identification marks
Stem prostrate to ascending, branched. Leaves alternate, trifoliate, leaflets obovoid, the middle one having a conspicuously longer petiole than the lateral ones. 20–50 flowers in a globular-ovoid head which is 7–15 mm wide. The petals remain on the flower after blossoming, dry and brownish-yellow (used for aerial distribution of seeds).

Habitat and distribution
Balks, dry meadows, railway embankments, paths, roadsides, fallow land. Found on soil which is rich in lime but low in nutrients. Found throughout Britain.

Additional information
Synonym: *T. procumbens*.

Pea family *Fabaceae (Leguminosae)*
Flowering time: May–Oct.
Height of growth: 5–25 cm
Dicotyledonous; Annual

Identification marks
Stem thin, round, prostrate to ascending. Leaves trifoliate, alternate, bluish-green. 10–25 flowers usually standing in ascending position and forming a globular head approx. 5 mm wide. Petals later become brown, persistent.

Habitat and distribution
Not too dry meadowland and pastures, wayside verges, river banks. Likes soil which contains nutrients and is relatively rich in lime. Common, but rarer in the north or north-west of Scotland.

Additional information
Synonyms: *T. filiforme*, *T. minus*.

Common Bird's-foot-trefoil
Lotus corniculatus

Greater Bird's-foot-trefoil
Lotus uliginosus

Pea family *Fabaceae (Leguminosae)*
Flowering time: June–Sept.
Height of growth: 10–40 cm
Dicotyledonous; Perennial

Identification marks
Stem ascending, usually filled with pith. Flowers 6–15 mm long, often tinged with red, 2–7 in cymose head. Leaves have three leaflets plus 2 leafy stipules at the base of the petiole; leaflets are obovoid, and may be glabrous, hairy or ciliate.

Habitat and distribution
Dry meadows, balks, moist meadows. Usually calcicolous. Throughout most areas.

Additional information
A plant with very many different forms varying in degree of hairiness, structure of the calyx and other features. Mountain plants are often dwarf, though similar to lowland forms in other ways.

Pea family *Fabaceae (Leguminosae)*
Flowering time: June–Aug.
Height of growth: 10–60 cm
Dicotyledonous; Perennial

Identification marks
Stem ascending to erect, with wide tubular hollow stem. Flowers 10–12 mm long, 5–12 of them in a cymose head. Before flowering the 5 calyx teeth are clearly folded back and the flower buds are usually tinged with red. Leaves are made up of 5 segments (3 'genuine' leaflets and the 2 similar stipules at the base of the stem), usually glabrous, bluish on the undersides.

Habitat and distribution
Damp meadows, ditches, river banks. On wet soils rich in nutrients and often low in lime. These soils may dry out in the summer. Throughout most areas, less common in the north.

Additional information
Synonym: *L. pedunculatus*. Very similar: see the entry on the left.

Dragon's-tooth
Tetragonolobus maritimus

Horseshoe Vetch
Hippocrepis comosa

Pea family *Fabaceae (Leguminosae)*
Flowering time: May–July
Height of growth: 10–25 cm
Dicotyledonous; Perennial

Identification marks
Stem prostrate to ascending, bluish-green like the leaves. Flowers solitary, 2–3 cm long, pale yellow. Leaves trifoliate with 2 large leafy stipules at the base, the lower leaves petiolate, somewhat fleshy, the upper ones with short petioles or sessile. Pods are long (up to 5 cm) and thin, with 4 wide longitudinal ribs.
Habitat and distribution
Rough grassland. Calcicolous. Tolerates salt. Local to rare in southern England (except the south-west) and a few other places.
Additional information
Is sometimes attributed to the Trefoil genus: *Lotus siliquosus* (see previous page).

Pea family *Fabaceae (Leguminosae)*
Flowering time: May–July
Height of growth: 10–40 cm
Dicotyledonous; Perennial

Identification marks
Stem prostrate to ascending. Leaves have long petioles, imparipinnate, with 9–15 oblong leaflets. Flowers approx. 1 cm long, 4–10 of them in a capitate head. Base of the petals narrows down conspicuously like a stem. Pods are divided into characteristic horseshoe-shaped segments.
Habitat and distribution
Poor turf, sunny balks, cliffs. Likes a warm calcareous soil. Scattered and local as far north as southern Scotland where suitable conditions exist.

Yellow Vetchling
Lathyrus aphaca

Meadow Vetchling
Lathyrus pratensis

Pea family *Fabaceae (Leguminosae)*
Flowering time: June–Aug.
Height of growth: 15–100 cm
Dicotyledonous; Annual

Identification marks
Stem prostrate, ascending or climbing. Leaves opposite, bluish green, heart-shaped to ovoid. Between each pair of 'leaves' is a long tendril (this is the remainder of the actual leaf, whereas the 'leaves' are in fact the much-enlarged stipules). Flowers are usually solitary, axillary, petiolate.
Habitat and distribution
Dry places on light soil. Likes calcareous loamy soil rich in nutrients and in a sunny location. Very local in south and south-east England extending to the Midlands and Wales.

Pea family *Fabaceae (Leguminosae)*
Flowering time: May–Aug.
Height of growth: 30–120 cm
Dicotyledonous; Perennial

Identification marks
Stem ascending or climbing, quadrangular. Leaves paired pinnate; consisting of 2 lanceolate leaflets and a simple or branched tendril. At the base of the stem 2 pointed leafy stipules. Flowers 1–1.5 cm long. Axillary racemes with long peduncles.
Habitat and distribution
Meadows, damp meadows, hedges, woodland margins. Likes a loamy soil which is not too dry and which is rich in humus and nutrients. Common throughout Britain.

Touch-me-not Balsam
Impatiens noli-tangere

Small Balsam
Impatiens parviflora

Balsam family *Balsaminaceae*
Flowering time: July–Sept.
Height of growth: 20–60 cm
Dicotyledonous; Annual

Identification marks
Stem erect, glossy; swollen at the nodes, branched towards the top. Leaves alternate, ovoid, coarsely serrate. 2–4 flowers in axillary racemes, hanging, 3–4 cm long, with curved spur; red spots on the inside. Fruit is a somewhat fleshy 5-lobed capsule; if it is touched when ripe it explodes violently (ejecting the seeds).
Habitat and distribution
Damp mixed and deciduous woodland, stream sides. Found on wet loamy soil rich in nutrients. Needs shade. Very local, occurring in many places though often only as a casual.
Additional information
Also called Yellow Balsam.

Balsam family *Balsaminaceae*
Flowering time: July–Nov.
Height of growth: 30–100 cm
Dicotyledonous; Annual

Identification marks
Stem erect, usually branched and leafy at the top; slightly swollen at the nodes. Leaves alternate, ovoid, toothed. 4–10 flowers in erect axillary racemes; approx. 1 cm long, pale yellow; spur straight. Fruit erect; if it is touched when ripe it explodes violently and ejects the seeds.
Habitat and distribution
Woods, thickets, rubble heaps, waste places. Found on well-saturated soils rich in nitrogen and low in lime. Likes shady locations. Local in south and east England but also found in other areas; usually in large numbers.

Wild Pansy (Heart's-ease)
Viola tricolor

Yellow Wood Violet
Viola biflora

Violet family *Violaceae*
Flowering time: April–Sept.
Height of growth: 10–20 cm
Dicotyledonous; Annual

Identification marks
Stem erect or ascending, usually branched. Leaves longer than wide, lower ones usually heart-shaped to ovoid, crenate. Stipules large, palmately lobed. Flowers solitary on long pedicels. Short spur; lower petals usually have blackish stripes on them.

Habitat and distribution
Fields, path balks, rubble heaps, dunes, mountain meadows. Mostly common but sometimes only local.

Additional information
Varies in habitat, petal length and colour. Several ssp: 3 in W. Europe. *Tricolor*, flowers bluish, rarely all yellow; throughout. *Curtisii*, low-growing, usually near sea. Flowers various colours. Throughout W. Europe. *Subalpina*, erect. Flowers yellow, sometimes upper petals violet. Mountains from S. and C. Europe.

Violet family *Violaceae*
Flowering time: May–Aug.
Height of growth: 8–15 cm
Dicotyledonous; Perennial

Identification marks
Stem ascending to erect. Large rosette leaves with long petioles; only 2–4 smaller alternate stem leaves: all heart to kidney-shaped, wider than they are long, crenate margin. Often 2 flowers on pedicels in leaf axil. Petals narrow, 4 pointing upwards, the lower one having a short spur and brown stripes on the front.

Habitat and distribution
Not British. Mountain and lowland forests, Alpine pastures, high shrub areas. Likes moist shady locations. Calcicolous. Apart from outlying areas in central Germany is found only in the Alps and their foothills. Scattered in these locations; up to 2,500 m. Mountains of central and southern Europe.

Wood Sage
Teucrium scorodonia

Large-flowered Hemp-nettle
Galeopsis speciosa

Mint family *Lamiaceae (Labiatae)*
Flowering time: July–Sept.
Height of growth: 15–30 cm
Dicotyledonous; Perennial

Identification marks
Stem erect, quadrangular, leaves opposite and decussate. Entire plant covered with soft hairs. Leaves have short petioles, ovoid, heart-shaped at the base, crenate. Flowers in pairs in a terminal raceme. Flowers approx. 1 cm long, greenish-yellow, without an upper lip. Lower lip has 3 lobes; middle lobe is large spoon-shaped.

Habitat and distribution
Open woodland, woodland paths, forest margins, heaths, thickets. Likes acid sandy ground low in nutrients and lime. Common throughout but avoids limestone regions.

Mint family *Lamiaceae (Labiatae)*
Flowering time: July–Sept.
Height of growth: 30–100 cm
Dicotyledonous; Annual

Identification marks
Stem erect, quadrangular, swollen at nodes and covered with stiff bristles, hairs bearing yellow glands, especially in the upper half of each internode. Leaves opposite, ovoid, toothed. Petioles 1–4 cm. Verticillasters arranged one above another. Flowers 2.5–4 cm long, pale yellow. Upper lip helmet-shaped. Middle lobe of the lower lip violet-coloured, rarely all yellow.

Habitat and distribution
Usually on cultivated land. Scattered throughout.

Additional information
Similar: Downy Hemp-nettle (*G. segetum*), nodes not swollen, downy, corolla pure yellow. In similar habitats in England and Wales but rare.

Yellow Archangel
Lamiastrum galeobdolon

Jupiter's Distaff
Salvia glutinosa

Mint family Lamiaceae (Labiatae)
Flowering time: May–July
Height of growth: 20–60 cm
Dicotyledonous; Perennial

Identification marks
Stem quadrangular, leaves opposite and decussate; sterile stems prostrate, flowering ones ascending or erect. Leaves similar to stinging nettles. Several verticillasters usually with 6 flowers. Lower lip with brown lines or markings.

Habitat and distribution
Woodland, especially coppiced areas. Only on moist soil containing mull and rich in nutrients. Common in England and Wales, rare in Scotland.

Additional information
Synonyms: *Galeobdolon luteum*, *Lamium galeobdolon*. Ssp. *flavidum*, with smaller flowers, in mountains of C. and S.E. Europe. Ssp. *montanum*, with sharply serrate leaves, more common in the S. Europe. Ssp. *galeobdolon*, more bluntly toothed leaves, more common in N. Europe.

Mint family Lamiaceae (Labiatae)
Flowering time: July–Oct.
Height of growth: 50–120 cm
Dicotyledonous; Perennial

Identification marks
Stem erect, bluntly quadrangular, usually glabrous at the bottom, at the top covered with glandular-sticky hairs. Leaves oblong-ovoid with spear-shaped base, petiolate to almost sessile, more or less glandular-sticky; margin is coarsely toothed. Flowers 3–5 cm long, (usually) 4–6 in each verticillaster; up to 16 verticillasters.

Habitat and distribution
Not British. Mountain, canyon and lowland forests, highshrub areas, bushy slopes. Likes moist loamy soil containing mull and rich in nutrients and usually in lime. Prefers semi-shady location. Only found in the Alps and outlying foreland. Scattered; hardly as high as 1,500 m. Mountains of central and southern Europe.

Common Cow-wheat
Melampyrum pratense

Small Cow-wheat
Melampyrum sylvaticum

Figwort family *Scrophulariaceae*
Flowering time: May–Oct.
Height of growth: 10–60 cm
Dicotyledonous; Annual.

Identification marks
Stem ascending to erect. Leaves opposite linear-lanceolate, rough. Unilateral lax spikes. Flowers 1–2 cm long, bracts green, upper ones toothed. Calyx glabrous.
Habitat and distribution
Open woodland, heaths, wet meadows. Found on soil which is always slightly acid, low in nutrients and usually in lime, and slightly dry to damp. Humus indicator. Common throughout.
Additional information
Very many different forms, to a large extent showing gradual variation from north to south in Europe. Also a number of habitat-related forms.

Figwort family *Scrophulariaceae*
Flowering time: June–Aug.
Height of growth: 5–35 cm
Dicotyledonous; Annual

Identification marks
Stem ascending or erect. Leaves opposite, lanceolate, almost glabrous. Lax unilateral spikes. Flowers up to 1 cm long, usually rich yellow. Bracts green. Calyx glabrous.
Habitat and distribution
Mossy woodland acid with humus, thickets and heaths. Found on loamy soils low in lime with surface acidity. Local to somewhat rare in mountainous parts of northern England and Scotland, reaching about 400 m.
Additional information
A number of habitat/seasonal forms are known, the main ones being spring, autumn and mountain forms.

Leafy Lousewort
Pedicularis foliosa

Moor-king
Pedicularis sceptrum-carolinum

Figwort family *Scrophulariaceae*
Flowering time: June–Aug.
Height of growth: 20–50 cm
Dicotyledonous; Perennial

Identification marks
Stem erect, unbranched, thick foliage towards the top. Leaves alternate, lobed; lobes are bipinnate. Flowers in a dense spike. Bracts are longer than flowers, similar to the stem leaves. Flowers 2–2.5 cm long, whitish-yellow; upper lip unspotted, covered with rough hairs.

Habitat and distribution
Not British. Alpine mats, streams, crook timber areas. Found on stony calcareous ground which is not too dry. Mountains in southern and southern-central Europe to Spain.

Additional information
Similar: *P. oederi*, bracts shorter than flowers, upper lip of flower glabrous, red spotted. Stony Alpine turf, pebbly ground. Rare. Mountains from Alps and Bulgaria north to the Arctic.

Figwort family *Scrophulariaceae*
Flowering time: June–Aug.
Height of growth: 30–100 cm
Dicotyledonous; Perennial

Identification marks
Stem erect, unbranched, little foliage, glabrous. Basal leaves numerous, pinnately divided with unpaired pinnate lobes. Stem leaves similar, smaller. Long, lax raceme with many flowers. Flowers pale yellow, 3–3.5 cm long; lower lip has red tip. Upper and lower lip come together so that the blossom remains closed.

Habitat and distribution
Not British. Wet meadows, river banks, communities in silted areas. Likes very wet peaty and loamy soil usually rich in lime. Very rare and numbers are on the decline. Only found in a few places in north and central Europe.

Greater Yellow Rattle
Rhinanthus alectorolophus

Yellow Rattle
Rhinanthus minor

Figwort family *Scrophulariaceae*
Flowering time: May–July
Height of growth: 10–60 cm
Dicotyledonous; Annual

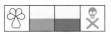

Identification marks
Stem erect, branched or not. Leaves opposite, oblong, sharply toothed. Axillary flowers, 2 cm long. Calyx wide, tufted hairs. Corolla tube slightly bent; upper lip has blue tooth.
Habitat and distribution
Meadows, balks, fields. Semi-parasite. Found on loamy soil rich in nutrients and often calcareous. Common in central Europe; up to over 2,000 m; rarer in the north.
Additional information
Not British. Many different forms. Different summer and autumn forms. Similar: Narrow-leaved Yellow-rattle, *R. angustifolius* (= *serotinus* = *major*): calyx glabrous, stem virtually glabrous to hairy. Meadows, fields, hedges. Scattered in Scotland and N. England, rarer in the south. Protected in Britain.

Figwort family *Scrophulariaceae*
Flowering time: May–Aug.
Height of growth: 10–40 cm
Dicotyledonous; Annual

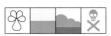

Identification marks
Stem erect, slightly hairy, usually with black spots. Leaves opposite, narrow-lanceolate, toothed. Flowers axillary; 1.5 cm long. Calyx glabrous but with hairy margins. Corolla tube straight. Upper-lip tooth white or light blue.
Habitat and distribution
Meadows, semi-dry turf. Semi-parasite; likes somewhat moist ground usually low in nutrients and lime. Common throughout.
Additional information
Very many different forms: not just lowland and mountain forms but also the respective early and (richly branched) late summer forms. Synonyms: *Alectorolophus minor*, *R. crista-galli*.

Large Yellow Foxglove
Digitalis grandiflora

Figwort family *Scrophulariaceae*
Flowering time: June–July
Height of growth: 60–120 cm
Dicotyledonous; Perennial

Identification marks
Stem erect, unbranched, at the top glandular-hairy. Leaves alternate, oblong-ovoid, margin ciliate and toothed. Unilateral terminal raceme. Flowers 3–4.5 cm long, campanulate to inflated, drooping, sulphur yellow, brownish spots on the inside; short upper lip, lower lip trilobed.

Habitat and distribution
Not British. Mountain forests, deciduous woodland, mixed woodland, thickets, glades; more rarely on mountain meadows and Alpine mats. Found on relatively nutritous soil, often free of lime, but with seepage water. Central Europe to northern Greece.

Small Yellow Foxglove
Digitalis lutea

Figwort family *Scrophulariaceae*
Flowering time: June–July
Height of growth: 40–80 cm
Dicotyledonous; Perennial

Identification marks
Stem erect, unbranched, glabrous. Leaves alternate, glabrous, oblong-ovoid. Unilateral raceme; slightly glandular. Flowers 2–2.5 cm long, tubular, pale yellow, unspotted on the inside. Upper lip 2 lobes, lower lips 3 lobes.

Habitat and distribution
Not British. Open woodland, sunny thickets, rocky heaths. Found on not too dry ground moderately rich in nutrients and usually calcareous. Very rare. Eastern and central Europe.

Additional information
Almost equally common is the hybrid with the Red Foxglove (see p.280): flowers 3–3.5 cm long, pale yellow, upperside slightly tinged with red; leaves have thin covering of short felt-like hairs (*D.* x *purpurascens*). Not British.

Common Toadflax
Linaria vulgaris

Greater Bladderwort
Utricularia vulgaris

Figwort family *Scrophulariaceae*
Flowering time: July–Oct.
Height of growth: 30–80 cm
Dicotyledonous; Perennial

Identification marks
Stem erect, usually unbranched, glabrous. Leaves alternate, sessile, linear-lanceolate, somewhat curled up along the edge. Dense terminal raceme. Flowers 1.5–2.5 cm long with a spur almost as long, pale yellow with orange palate.

Habitat and distribution
Weedy areas, paths, walls, fences, railway ballast, waste places, fields, clearings. Likes warmth. Found on loose, not too dry soil which is rich in nutrients and minerals. Needs light. Common in England and Scotland.

Additional information
Varies especially in the colour of the flowers.

Butterwort family *Lentibulariaceae*
Flowering time: July–Aug.
Height of growth: 15–45 cm
Dicotyledonous; Perennial

Identification marks
Free-floating. Leaves divided into hair-like segments, with many (20–200) small bladders 2–4 mm wide (to trap small insects). Leaf segments slightly ciliate. Lax raceme above the water level. Flowers golden-yellow, up to 2 cm long.

Habitat and distribution
Stagnant or slow-flowing warm water, low in lime but containing nutrients. Local throughout Britain.

Additional information
Very similar: *U. australis* (= *neglecta*), flowers pale yellow; England, Wales, southern Scotland, rare in the east. Other similar species have stems of two kinds, one bearing normal green leaves, the other colourless with reduced leaves and buried in the ground. On some species only these anchoring stems bear bladders.

Mistletoe
Viscum album

Box
Buxus sempervirens

Mistletoe family *Loranthaceae*
Flowering time: Feb.–April
Height of growth: 20–100 cm
Dicotyledonous

Identification marks
Parasite living on trees. Bushy shrub, richly forked branches; whole plant yellowish-green. Leaves opposite, leathery, evergreen, oblong-ovoid, narrower towards base. Flowers in dense cymes, inconspicuous. Plant dioecious. Berries globular, white to yellowish-coloured.

Habitat and distribution
Orchards, undergrowth along streams, woodland, also on individual trees. Most common on apples, rarely on conifers. Common in S. England and parts of the Midlands, rarer elsewhere, absent in Scotland.

Additional information
3 ssp: *Album*, on apple trees, poplars and other deciduous trees, *Abietis*, on fir trees, and *Austriacum (= laxum)*, on pines and larches. C. and S. Europe.

Box family *Buxaceae*
Flowering time: April–May
Height of growth: 2–5 m
Dicotyledonous

Identification marks
Low shrub, more rarely small tree. Richly branched. Branches short, usually erect. Leaves approx. 2 cm long, opposite, leathery, tough, evergreen, oblong to elliptical; petioles short. Flowers in axillary clusters; unisexual. Terminal female flower (often more than 4 sepals) forms a cluster with many male flowers (4 sepals).

Habitat and distribution
On sunny bushy slopes and in beech woodland on chalk or limestone soils. Very local though sometimes abundant in southern England. Frequently planted elsewhere and very often becoming naturalized.

Additional information
Ornamental shrub with many variations of growth, leaf shape and leaf colour.

Buckthorn
Rhamnus catharticus

Cornelian Cherry
Cornus mas

Buckthorn family *Rhamnaceae*
Flowering time: May–June
Height of growth: 4–6 m
Dicotyledonous

Identification marks
Branches and leaves opposite. Leaves petiolate, ovoid; margin finely toothed. Many small branches terminating in pointed thorn. Flowers in sparse axillary clusters on the previous year's growth, yellowish-green, inconspicuous, pleasant smelling. Black berries, pea-sized, inedible (or have purging effect).

Habitat and distribution
Hedges, dry thickets, oak and ash woods. Prefers calcareous soils. Scattered in England and Wales except for S.W. areas, and Scotland.

Additional information
Similar: Rock Buckthorn, *R. saxatilis*, more delicate in every respect; leaves only 1–1.5 cm long. The Alder Buckthorn, *Frangula alnus* (see p.379), has no thorns and leaves are alternate.

Dogwood family *Cornaceae*
Flowering time: Feb–March
Height of growth: 2–8 m
Dicotyledonous

Identification marks
Flowers in small lateral umbellate clusters, appearing before the leaves. Leaves deciduous, opposite, ovoid to elliptical, margin entire; approx. 10 cm long; petiole approx. 1 cm long; on the underside in the angles of the curved veins there are tufts of hair ('mite nests'). Red oblong drupe, drooping, approx. 1 cm long; edible, some-what acid taste.

Habitat and distribution
Calcicolous, likes warmth. Often planted as a garden ornamental, only rarely escaping and growing wild.

Field Maple
Acer campestre

Norway Maple
Acer platanoides

Maple family *Aceraceae*
Flowering time: May–June
Height of growth: 2–20 m
Dicotyledonous

Identification marks
Small tree or (often) only 2–3 m high shrub. Older bark shows reticulate fissures, young (4–5 years old) twigs often have cork ribs forming wing-like extensions. Leaves opposite, petiolate, 5-lobed, these are in turn bluntly lobed. Flowers in erect corymbose panicles; usually appearing with or just after the foliage. Fruit a pair of winged samaras.

Habitat and distribution
Deciduous woodland, thickets, hedges. On soil rich in nutrients and minerals. Common in southerly and central parts of England, especially on basic soils, rarer elsewhere.

Maple family *Aceraceae*
Flowering time: April–May
Height of growth: 20–35 m
Dicotyledonous

Identification marks
Moderately large tree with wide to ovoid crown. Older bark has numerous short fissures; young twigs glabrous, shiny brown. Leaves opposite, petiolate, 5–15 cm long, 5-lobed; lobes are long-toothed. Flowers in corymbose panicles, usually appearing shortly before the foliage. Fruit a pair of winged samaras.

Habitat and distribution
Woods and hedges. Likes damp loose soil often containing pebbles. Fairly common in most areas though often only planted.

Additional information
There are many other similar maple species, which are particularly well-suited as ornamental trees. The Sycamore (see p.381) is the more common species.

Small-leaved Lime
Tilia cordata

Red-berried Elder
Sambucus racemosa

Lime-tree family *Tiliaceae*
Flowering time: June
Height of growth: 10–25 m
Dicotyledonous

Identification marks
Tree with spreading crown. Older bark somewhat dark with fine longitudinal fissures; young twigs reddish-brown to olive green. Leaves alternate, petiolate, crookedly heart-shaped, unevenly serrate; underside glabrous, apart from rust-yellow tufts of hair in the angles of the veins. 4–10 flowers in pendulous cyme; on main stem is a tongue-shaped membraneous covering leaf (flight organ).
Habitat and distribution
Variety of soils, especially common in woods on limestone cliffs. Scattered throughout England and Wales, planted in N. England and Scotland; also often along avenues and in parks.
Additional information
Similar: Large-leaved Lime (*T. platyphyllos*). Leaves larger, white tufts of hair in angles of veins.

Honeysuckle family *Caprifoliaceae*
Flowering time: April–May
Height of growth: 5–15 cm
Dicotyledonous

Identification marks
Shrub, very rarely a small tree. Twigs have dark brown bark with coarse pores and brownish-red pith. Leaves opposite, short petioles, pinnate; 3–7 leaflets (usually 5), oblong-ovoid, pointed, margin serrate. Flowers in erect, dense, ovoid panicles, strong smelling. Berries globular, red.
Habitat and distribution
Likes somewhat stony ground free of lime. Widely planted, sometimes escaping and becoming naturalized.

Barberry
Berberis vulgaris

Gorse
Ulex europaeus

Barberry family *Berberidaceae*
Flowering time: May–June
Height of growth: 1–2.5m
Dicotyledonous

Identification marks
Thorny, richly branched shrub. Branches rod-like, often tinged with red. Bark pale grey, wood yellow, hard. Leaves in clusters on the short lateral shoots, ovoid, finely spiny-toothed; short petioles. On the long shoots the leaves are clearly alternate, with trifid thorns at the base. Pendulous lateral racemes with 6–12 flowers approx. 1 cm long. Berries elongate-cylindrical, scarlet in colour with bitter taste.

Habitat and distribution
Open woodland, hedgerows, thickets. On warm calcareous soil rich in nutrients. Scattered throughout Britain but nowhere in large numbers. Planted for ornament and the fruit which has edible pulp (not the stone which is poisonous).

Additional information
Found in many different species.

Pea family *Fabaceae (Leguminosae)*
Flowering time: May–June
Height of growth: 0.5–2 m
Dicotyledonous

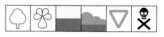

Identification marks
Very prickly shrub. Stem erect, twigs green, grooved. Leaves of older plants become needle-shaped spines: simple or trifid, the upper ones linear stiff and sharply pointed. 1–3 flowers in the upper leaf axils, 1.5–2 cm long, scented.

Habitat and distribution
Rough grassland, open dry woodland, heaths. Somewhat calcicolous, does not withstand frost. Throughout Britain, but planted in parts of the far north.

Additional information
Also called Furze or Whin.

German Greenweed
Genista germanica

Dyer's Greenweed
Genista tinctoria

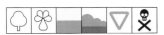

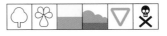

Pea family *Fabaceae (Leguminosae)*
Flowering time: May–June
Height of growth: 20–50 cm
Dicotyledonous

Identification marks
Half-shrub; woody at the bottom, young shoots herbaceous. Ascending or erect, branched at the top, thorny at the bottom. Leaves simple, with rough hairs, elongate-elliptical, 1–2 cm long. Terminal racemes. Flowers 1 cm long, hairy.

Habitat and distribution
Not British. Heaths, open woodland, path balks. Calcifugous; likes warmth. Scattered, hardly found over 700 m; central and southern Europe.

Additional information
Similar: Petty Whin (*G. anglica*), also thorny, but entire plant glabrous, bluish-green; heaths and moors on soil low in lime: scattered throughout.

Pea family *Fabaceae (Leguminosae)*
Flowering time: July–Sept
Height of growth: 30–70 cm
Dicotyledonous

Identification marks
Shrub: woody at the bottom, young shoots herbaceous. Prostrate to ascending, more richly branched at the top. Stem and branches do not have thorns. Leaves oblong-lanceolate, 1–3 cm long, with soft hairs along the margin. Flowers axillary, around 1.5 cm long, glabrous.

Habitat and distribution
Rough turf and grassy places. Found on loamy soil which is not too dry and preferably low in nutrients. Able to withstand lime. Throughout England and Wales, rare in southern Scotland and absent in the north.

Additional information
Similar: Hairy Greenweed (*G. pilosa*); also without thorns, but flowers covered with silky hairs, 1–2 in leaf axils. Cliffs and sandy heaths; calcifugous. Rare in S. England and Wales.

Black Broom
Lembotropis nigricans

Broom
Cytisus scoparius

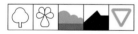

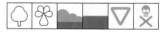

Pea family *Fabaceae (Leguminosae)*
Flowering time: June–Aug.
Height of growth: 50–150 cm
Dicotyledonous

Identification marks
Shrub with ascending to erect twigs with brown bark; erect green flower buds, round and finely grooved. Leaves alternate, all made up of 3 leaflets which are ovoid, approx. 2 cm long. Leaves, petioles and twigs covered with short hairs. Erect terminal raceme, 10–30 cm long, without foliage. Flowers around 1 cm long.

Habitat and distribution
Not British. Open woodland, forest margins, thickets, path balks, rocks. Likes warmth. Found on shallow stony or sandy ground. Rare. Southern Germany, southern Alps to eastern Europe.

Additional information
Synonym: *Cytisus nigricans*.

Pea family *Fabaceae (Leguminosae)*
Flowering time: June–Aug.
Height of growth: 50–200 cm
Dicotyledonous

Identification marks
Shrub with old twigs which usually stand awry; brown bark; rod-like young twigs green in colour and angular. Leaves alternate, falling off rapidly; the lower ones have 3 leaflets which are each approx. 1 cm long, the upper ones are sessile and undivided, elongate-ovoid, flowers approx. 2 cm long, solitary or in twos, short pedicels, in the axils of the upper leaves.

Habitat and distribution
Woodland margins, glades, clearings, heaths, path balks, waste places. Calcifugous. Often planted to stop soil erosion and to enrich the soil with nitrogen. Not resistant to frost. Scattered throughout Britain.

Additional information
Synonym: *Sarothamnus scoparius*.

Shrubby Milkwort
Polygala chamaebuxus

Fly Honeysuckle
Lonicera xylosteum

Milkwort family *Polygalaceae*
Flowering time: April–June
Height of growth: 10–20 cm
Dicotyledonous

Identification marks
Prostrate dwarf shrub with ascending branches. Leaves leathery, evergreen, lanceolate to elliptical, margin entire, the lower leaves ovoid, smaller than the upper ones. 1–2 flowers in the leaf axils, 12–15 mm long, yellow, often brown-reddish tinged. Racemes have few flowers, interspersed with leaves.

Habitat and distribution
Not British. Dry pine forests; thickets, heaths, poor stony turf, rocks. On sunny stony ground rich in lime and low in nutrients. Must be dry at least part of the year. Rare; mountains of central and southern Europe.

Honeysuckle family *Caprifoliaceae*
Flowering time: May–June
Height of growth: 1–2 m
Dicotyledonous

Identification marks
Branched shrub with rod-like hollow twigs. Flowers always in pairs on a stem, the lower globular ovaries having grown together. Corolla hairy. Leaves opposite, simple, wide ovoid, blunt, margin entire. Shiny red (double) berries.

Habitat and distribution
Deciduous woodland, in hedges. Calcicolous. Needs loose soil rich in humus and nutrients. Scattered in England and Wales, and a few places in Scotland. Mostly introduced.

Additional information
The flowers are never pure white. Often the very similar *L. ruprechtiana* from China is planted in parks and gardens. Flowers do not have hairs on the outside, snow-white in colour, changing to yellow.

Water-plantain
Alisma plantago-aquatica

Water-pepper
Polygonum hydropiper

Water Plantain family *Alismataceae*
Flowering time: June–Aug.
Height of growth: 20–100 cm
Monocotyledonous; Perennial

Identification marks
Flowers in whorls, their parts (petals etc.) in threes, the petals withering very rapidly, white or lilac, yellowish base. Leaves oval, robust, on long stalks, forming basal rosette.

Habitat and distribution
Banks of stagnant or slow-flowing waters, in reed beds and sedge, also in ditches. Throughout most of Britain, rarer in the north. Indicator of muddy ground rich in nutrients.

Additional information
Similar to some closely related species. They are distinguished by their leaves and their habitat: Narrow-leaved Water-plantain (*A. lanceolatum*), leaves narrow; Ribbon-leaved Water-plantain (*A. gramineum*), leaves ribbon-like, submerged, protected in Britain. Both aquatic plants.

Dock family *Polygonaceae*
Flowering time: July–Oct.
Height of growth: 25–75 cm
Dicotyledonous; Annual

Identification marks
Spike lax, thin and slender. Stem erect. Leaves elongate-lanceolate, narrower towards the base and the tip. Flowers greenish or reddish.

Habitat and distribution
Ditches, river banks, damp pathways. Prefers ground low in lime. Withstands temporary flooding. Indicator of nitrogen. Common throughout Britain except northern Scotland.

Additional information
The plant may be confused with Tasteless Water-pepper (*P. mite*). When chewed this does not taste peppery and sharp. It occurs in the same habitats, but is rarer.

Common Sorrel
Rumex acetosa

Sheep's Sorrel
Rumex acetosella

Dock family *Polygonaceae*
Flowering time: May–June
Height of growth: 30–100 cm
Dicotyledonous; Perennial

Identification marks
Leaves tough and thickish, spear-shaped at the base with backward-pointing lobes. Upper leaves more or less sessile and clasping the stem. Leaves have a somewhat acid taste. Flower panicles slender. Plant dioecious.

Habitat and distribution
Meadows and open woodland, on loamy soil rich in nutrients. Common throughout.

Additional information
This plant is eaten because of its high Vitamin C content. It contains oxalic acid. Large quantities are damaging to one's health. The similar Sheep's Sorrel (*R. acetosella*) rarely grows higher than 30 cm and can be distinguished easily from Common Sorrel by the leaf-lobes and upper leaves, which do not clasp the stem.

Dock family *Polygonaceae*
Flowering time: May–Aug.
Height of growth: 8–30 cm
Dicotyledonous; Perennial

Identification marks
Plant is smaller than 30 cm. When chewed the leaves have a distinctly bitter taste. All leaves petiolate, the upper ones not clasping the stem. Leaves narrow, often with a spear-shaped base, the lobes forward-pointing.

Habitat and distribution
Dry meadows, sandy ground and heaths, more rarely on sandy fields. Indicator of sandy and acid soil. Common.

Additional information
Various races can be distinguished within the species. On very poor sandy soils is the similar *R. tenuifolius*, with narrower leaves, with inrolled margins.

Common Poppy
Papaver rhoeas

Cuckoo Flower
Cardamine pratensis

Poppy family *Papaveraceae*
Flowering time: June–Aug.
Height of growth: 20–60 cm
Dicotyledonous; Annual – Biennial

Identification marks
Pedicels have projecting bristles. Ovary and capsule glabrous, the capsule more or less globular. Plant contains whitish milky juice (latex). Leaves deeply pinnate, serrate.

Habitat and distribution
Cornfields, rubble heaps, roads and wayside places. Prefers calcareous loamy soil. Common in the south, becoming rarer in the north of Scotland.

Additional information
The similar Long-headed Poppy (*P. dubium*) has a club-shaped capsule. It occurs in the same localities as the Common Poppy but prefers soil low in lime. Both species have been reduced in numbers over recent years as a result of weed-killing through herbicides.

Mustard family *Brassicaceae (Cruciferae)*
Flowering time: April–June
Height of growth: 15–60 cm
Dicotyledonous; Perennial

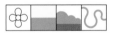

Identification marks
Basal leaves in rosette arrangement, pinnate, leaflets round, the terminal leaflet usually larger. Stem leaves pinnate with narrow segments. Stem hollow. Flowers in racemes, large (approx. 1–2 cm wide). Ovaries and fruit much longer than wide. Flower colour is very variable ranging from white, pink, lilac to deep blue-violet.

Habitat and distribution
Damp meadows and woodland on loamy soil. Common throughout.

Additional information
Similar: Narrow-leaved Bitter-cress (*C. impatiens*); flowers about 0.5 cm wide, whitish. Local in woodland in the west of Britain.

Coralroot
Cardamine bulbifera

Sea Rocket
Cakile maritima

Mustard family *Brassicaceae (Cruciferae)*
Flowering time: April–May
Height of growth: 30–70 cm
Dicotyledonous; Perennial

Identification marks
Flowers violet-coloured, usually tending more towards blue than to red. No rosette of leaves. Leaves at least partially pinnate. Upper leaves narrow and undivided. In the axils there are blackish-coloured bulbils.

Habitat and distribution
Deciduous and mixed woodland on soil rich in nutrients and usually calcareous. Local in southern England, Midlands to southern Scotland.

Additional information
The Coralroot reproduces mainly by means of its bulbils. The formation of seeds is rare or totally absent. Ants carry off the bulbils, which explains why there are often only small numbers of the plant in the localities where it occurs. Synonym: *Dentaria bulbifera.*

Mustard family *Brassicaceae (Cruciferae)*
Flowering time: June–Aug.
Height of growth: 15–45 cm
Dicotyledonous; Annual

Identification marks
Plant tinged with bluish-green. Stem richly branched, ascending. Leaves fleshy, undivided or pinnately divided, alternate. Flowers sweet-smelling, a good 0.5 cm long; dense racemes. Fruit in two sections, the lower one chisel-shaped, the upper one oval-shaped at the joining point with spear-like humps.

Habitat and distribution
All around the coasts of Britain, on sandy and shingle beaches and in dunes. Withstands high salt concentration, but does not need it to grow. Rarely carried inland and does not grow well there.

Great Burnet
Sanguisorba officinalis

Pale Willowherb
Epilobium roseum

Rose family *Rosaceae*
Flowering time: June–Sept.
Height of growth: 30–100 cm
Dicotyledonous; Perennial

Identification marks
Dense terminal heads of deep purplish-red or brownish-red flowers. Stem erect and branched in the upper part. Leaves large, pinnate. Leaflets petiolate, ovoid, margin crenate.

Habitat and distribution
Moist meadows. Grows on very damp, often peaty soil but also loamy. Does not form very large numbers. Common locally, mainly in western parts of Britain.

Additional information
The inconspicuous flowers are generally bisexual, i.e. they have both stamens and ovaries. Occasionally however you may find unisexual flowers. The very light seeds of the Great Burnet are dispersed by the wind.

Willow Herb family *Onagraceae*
Flowering time: July–Aug.
Height of growth: 25–60 cm
Dicotyledonous; Perennial

Identification marks
Many-flowered inflorescence. Petals 4–5 mm long, deeply notched, initially almost white, later streaked with pink. Stem glabrous at the bottom, further up slightly hairy. Leaves elongate-lanceolate, fairly long petioles, finely toothed.

Habitat and distribution
Woods, river banks, also in ditches. Likes calcareous soil rich in nutrients and wet from seepage. Throughout most of Britain but absent from mountainous areas.

Additional information
Similar: *E. alpestre*, petals 6–12 mm long, red right from initial flowering. Leaves in threes on lower part of stem, sessile, very clearly toothed. Scattered. Mountains of central and southern Europe.

Rosebay Willowherb
Epilobium angustifolium

Alpine Fireweed
Epilobium fleischeri

Willowherb family *Onagraceae*
Flowering time: July–Sept.
Height of growth: 30–120 cm
Dicotyledonous; Perennial

Identification marks
All leaves alternate, 1–2 cm wide and 5–15 cm long. Leaf uppersides dark green, undersides bluish-green. Flowers 2–3 cm in diameter. Style curved with 4 clear stigmas.

Habitat and distribution
Woodland, old ballast areas, e.g. railway tracks which have been shut down or are used only infrequently. Grows on loose nitrogenous soil. Common throughout though less so in the north.

Additional information
This plant is sometimes placed in a genus of its own (Fireweed – *Chamaenerion*).

Willowherb family *Onagraceae*
Flowering time: July–Aug.
Height of growth: 10–30 cm
Dicotyledonous; Perennial

Identification marks
Stem not erect but prostrate or ascending. Leaves alternate, only 1–3 mm wide. Flowers vivid red.

Habitat and distribution
Not British. Only found in central and southern Europe. There it grows on loose gravel and scree. Scattered, sometimes absent in a whole area.

Additional information
Hybrids between this species and most of the other species in the family have not been found. The Alpine Fireweed is also sometimes placed in the same genus as Fireweed (*Chamaenerion*), as is Rosemary Willowherb (*E. dodonaei*), which also has narrow leaves but an upright stem.

Great Willowherb
Epilobium hirsutum

Hoary Willowherb
Epilobium parviflorum

Willowherb family *Onagraceae*
Flowering time: May–Aug.
Height of growth: 80–150 cm
Dicotyledonous; Perennial

Identification marks
Flowers in corymbose racemes at the ends of the stem and main branches. Flowers around 2 cm in diameter. Petals notched, light purplish-red. Middle and lower leaves are 5–12 cm long. Lower leaves mostly opposite and sessile and slightly decurrent. Stem is hairy lower down.

Habitat and distribution
Ditches, reeds, damp woodland, fens. Prefers calcareous soils, withstands or requires occasional flooding. Common in all but the far north-west.

Additional information
Great and Hoary Willowherb are very similar but the Great Willowherb is decidedly the larger. Its flowers grow to twice the size of the Hoary Willowherb (*E. parviflorum*).

Willowherb family *Onagraceae*
Flowering time: June–Sept.
Height of growth: 15–80 cm
Dicotyledonous; Perennial

Identification marks
Whole stem has projecting hairs, stem usually erect, more rarely only ascending. Leaves elongate or lanceolate, barely over 7 cm long, the lower and middle ones opposite, the upper ones alternate. Flowers fairly large, approx. 1 cm in diameter. Petals deeply notched.

Habitat and distribution
Reed beds, ditches, moist woodland. Likes damp loamy soil rich in nutrients.

Additional information
As the Hoary Willowherb is often found in the same locations as the Great Willowherb, it used to be thought that these were two sexually different examples of the same species.

Broad-leaved Willowherb
Epilobium montanum

Water Mint
Mentha aquatica

Willowherb family *Onagraceae*
Flowering time: June–Aug.
Height of growth: 20–60 cm
Dicotyledonous; Perennial

Identification marks
Petals 8–10 mm long, pale pink to almost white, deeply notched. Stem usually only branched in the top part, also not too many standing close together. Plant does not, therefore, make a compact effect. Leaves in the lower half of the plant opposite, in the upper half sometimes in whorls of three, 4–8 cm long.

Habitat and distribution
Woodland, hedgerows, occasionally also in gardens. Likes moist somewhat stony ground. Frequently found on calcareous soil. Common throughout Britain.

Mint family *Lamiaceae (Labiatae)*
Flowering time: July–Oct.
Height of growth: 15–90 cm
Dicotyledonous; Perennial

Identification marks
Flowers grow in rounded terminal heads or in verticillasters in the axils of the upper leaves. The 'lip blossom' is so inconspicuous on the Water Mint that for the layman the plant does not even seem to be bilaterally symmetrical. It is taken as having 4 lobes. Stem 4-angled; leaves opposite and decussate, toothed.

Habitat and distribution
Reeds, river banks, ditches, wet meadows and fields. Common throughout Britain.

Additional information
The various types of Mint are hard to distinguish from one another because there are many hybrids. The best-known one is the Peppermint (*M. x piperita*). This hybrid is a cross between the Water Mint and the Spear Mint.

Knotgrass
Polygonum aviculare

Redshank
Polygonum persicaria

Dock family *Polygonaceae*
Flowering time: July–Oct.
Height of growth: 10–200 cm
Dicotyledonous; Annual

Identification marks
Stem prostrate. Flowers in axillary clusters, pale greenish-white, often tinged with pink.

Habitat and distribution
Weedy places on fields and along paths and waste places. Nitrogen indicator. Very common throughout. Resistant to trampling.

Additional information
Knotgrass can be clearly distinguished from other species in this genus by its prostrate, non-twining stem and the axillary flower clusters.

Dock family *Polygonaceae*
Flowering time: June–Oct.
Height of growth: 25–75 cm
Dicotyledonous; Annual

Identification marks
Stem somewhat branched, each branch terminating in a spike. Petiole projecting from a papery sheath (ochra) surrounding the stem at each node. Ochra fringed along margin. Leaves between petiole and leaf centre often somewhat hairy.

Habitat and distribution
Fields, rubble heaps, ditches. Likes nitrogen. Common throughout Britain.

Additional information
Similar: Pale Persicaria (*P. lapathifolium*). Leaf sheaths not fringed. Fields, ditches. Likes nitrogen.

Common Bistort
Polygonum bistorta

Wall Gipsy Weed
Gypsophila muralis

Dock family *Polygonaceae*
Flowering time: June–Aug.
Height of growth: 25–50 cm
Dicotyledonous; Perennial

Identification marks
Stem unbranched, bearing only a single spike at the end. The flowers on the spike are usually dense and thus create a compact cylindrical effect when the flowers are open. Leaves ovoid, greyish-green underneath. Rootstock twisted like a snake (hence the name).

Habitat and distribution
Damp meadows, damp woodland, occasionally on roadsides. Moisture indicator. Common. Usually forms large groups.

Additional information
Especially on damp meadows Common Bistort may form large colonies. Then it becomes a nuisance. In hay it does not dry well. Drainage and liming help to reduce numbers.

Pink family *Caryophyllaceae*
Flowering time: June–Oct.
Height of growth: 5–20 cm
Dicotyledonous; Annual

Identification marks
Flowers only 6–10 mm in diameter, pink, with darker veins. Calyx has dry membraneous stripes, toothed, teeth only approx. 1/3 of length of calyx. Stem ascending or stiffly erect, with forked branching. Leaves opposite, narrow linear, barely 1 mm wide.

Habitat and distribution
Most of Europe. In Britain only as a rare casual.

Additional information
Similar: Creeping Gypsophila (*G. repens*). Not British. Perennial plant. Flowers white, at most tinged with purple. Limestone scree and damp rubble heaps. Mountains of central and southern Europe. Creeping Gypsophila dams up scree.

Deptford Pink
Dianthus armeria

Carthusian Pink
Dianthus carthusianorum

Pink family *Caryophyllaceae*
Flowering time: July–Aug.
Height of growth: 30–60 cm
Dicotyledonous; Annual – Biennial

Identification marks
2–10 flowers in terminal cluster surrounded by narrow, densely packed bracts. Flowers approx. 1 cm in diameter. Epicalyx of 2 scales. Uppersides of the petals light purplish-red to dark pink, with numerous white spots; towards the throat of the corolla there are individual darker spots and individual hairs.
Habitat and distribution
Dry turf, hedgerows. Calcifugous. Rare, in lowland habitats in England, Wales and a few places in Scotland.
Additional information
Sweet-William (*D. barbatus*), epicalyx of 4 scales. Flowers larger. Calcareous soil. Popular garden plant sometimes found growing wild.

Pink family *Caryophyllaceae*
Flowering time: June–Sept.
Height of growth: 10–50 cm
Dicotyledonous; Annual – Perennial

Identification marks
4–10 flowers in a terminal cluster, but usually only one or two flowers blossom at a time. Petals dark red or deep pinkish-red. Calyx glabrous with hairy pointed epicalyx scales at the base, approx. 1/2 length of the calyx. Leaf sheaths on the stem at least twice as long as the leaves are wide.
Habitat and distribution
Dry turf, open woodland and dry thickets. Scattered.
Additional information
The Carthusian Pink is a rare escape from cultivation in Britain.

Maiden Pink
Dianthus deltoides

Cheddar Pink
Dianthus gratianopolitanus

Pink family *Caryophyllaceae*
Flowering time: June–Sept.
Height of growth: 15–45 cm
Dicotyledonous; Perennial

Identification marks
Stem erect or ascending, covered with rough, short hairs, branched at the top. Petals purplish-red, with clear white spots and relatively long white hairs. A dark line running parallel to the outer petal margin is usually clearly visible. Calyx cylindrical, glabrous, at the base with 2 epicalyx scales which are only half as long.

Habitat and distribution
Pastures, poor turf. Calcifugous. Prefers sandy loose soils low in nutrients. Absent in limestone regions, otherwise local throughout.

Additional information
Differs from the similar *D. seguieri* which always has a glabrous (smooth) stem.

Pink family *Caryophyllaceae*
Flowering time: June–July
Height of growth: 10–25 cm
Dicotyledonous; Perennial

Identification marks
Solitary flowers at the end of a stem, 2–2.5 cm in diameter. Petals along the outer edge have fairly short teeth. Calyx often tinged with violet, surrounded below by 4–6 epicalyx scales barely 1/4 as long as the calyx and which may be blunt or pointed. Leaves never have keel, bluish-green. Plant often forms small cushions.

Habitat and distribution
Very rare. Nowadays only found on the limestone cliffs of Cheddar Gorge.

Dianthus seguieri

Dianthus superbus

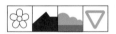

Pink family *Caryophyllaceae*
Flowering time: June–Aug.
Height of growth: 30–60 cm
Dicotyledonous; Perennial

Identification marks
1–8 flowers in loose head. They are deep pink in colour and often white-spotted near the base. The margins are bearded. Calyx cylindrical, glabrous, with 2–6 oval epicalyx scales at the base. Their pointed apices do not reach the edge of the calyx.

Habitat and distribution
Not British. Poor mats, acid mountain meadows, chestnut thickets. Alpine plant. Found in south-western Europe and western central Europe. Very rare.

Pink family *Caryophyllaceae*
Flowering time: June–Aug.
Height of growth: 30–90 cm
Dicotyledonous; Perennial

Identification marks
Flowers solitary, on pedicels. Petals lilac-coloured to deep pink, often with a strong tendency towards lilac. Petals divided beyond the middle, often with dark spots, bearded. Flower spread out, generally over 2.5 cm in diameter. Short epicalyx scales at the base of the calyx, these reaching no more than 1/3 of calyx length. Flowers usually have strong smell.

Habitat and distribution
Not British. Woodland thickets, mountain meadows, wet meadows. Scattered. Most of continental Europe.

Additional information
Similar: Pink (*D. plumarius*), flower spread out, under 2.5 cm in diameter, pink to white. Calyx scales reach to approx 1/4 the length of the calyx. Not British. Eastern central Europe. Rare.

Moss Campion
Silene acaulis

Red Campion
Silene dioica

Pink family *Caryophyllaceae*
Flowering time: July–Aug.
Height of growth: 1–10 cm
Dicotyledonous; Perennial

Identification marks
Plant grows in dense cushions. Flowers solitary on the stems which are usually very short at first but lengthen later. Petals purple to pink. Leaves opposite, upper ones ovoid, lower ones narrow, 5–15 mm long.
Habitat and distribution
Cliffs, ledges and screes in the mountains. Likes calcareous loose soils. North Wales, Lake district, Scotland.

Pink family *Caryophyllaceae*
Flowering time: May–June
Height of growth: 30–90 cm
Dicotyledonous; Perennial

Identification marks
Flowers numerous, in terminal dichasia. They are dull pink to purplish-red, rarely white. Plant is dioecious: male and female flowers are on separate plants. Female flowers have a calyx with 20 veins, male flowers have one with 10 veins, their flowers are usually smaller and duller. Entire plant covered with soft hairs. Leaves opposite, upper ones ovoid.
Habitat and distribution
Damp meadows, hedgerows and woodland. Common in most areas. Wetness indicator.
Additional information
Red Campion has been placed with other species in the genus *Melandrium: M. dioicum, M. rubrum.*

Sticky Catchfly
Lychnis viscaria

Ragged-Robin
Lychnis flos-cuculi

Pink family *Caryophyllaceae*
Flowering time: June–Aug.
Height of growth: 30–60 cm
Dicotyledonous; Perennial

Identification marks
Flowers grow in a fairly loose panicle. They reach a diameter of 2 cm. The petals are purplish-red and slightly notched at the apex. The dark viscid rings below the upper nodes on the stem are striking. Leaves opposite. Plant glabrous or only with slight covering of short hairs.
Habitat and distribution
Dry meadows, heaths, dry woodland. Prefers soil low in lime. Rare.
Additional information
The Sticky Catchfly can be easily recognized by its viscid rings. There is no proven knowledge regarding their importance but it is probably safe to assume that they ward off harmful insects.

Pink family *Caryophyllaceae*
Flowering time: May–June
Height of growth: 30–75 cm
Dicotyledonous; Perennial

Identification marks
The flowers grow in loose dichasia. They are either light pink or occasionally white. The petals are conspicuously deeply divided. Leaves opposite, spatulate to lanceolate.
Habitat and distribution
Meadows. Wetness indicator. Common.
Additional information
The Ragged-Robin is one of the plants which is frequented particularly often by leaf hoppers when laying their eggs. These larvae take what they need to survive from the juice of the plant and the combination of the juice plus their breath forms a saliva-like froth which remains on the plant. In the past it was not possible to explain the occurrence of this froth. It was thought to be 'devil's work'.

Soapwort
Saponaria officinalis

Saponaria ocymoides

Pink family *Caryophyllaceae*
Flowering time: July–Sept.
Height of growth: 30–90 cm
Dicotyledonous; Perennial

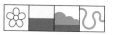

Identification marks
Stem erect, often tinged with red; leaves on stem are sessile, opposite, broadly ovoid. Flowers in dense terminal corymbs on the main stem and its branches. Petals spread out flat from long cylindrical calyx, only slightly crenate at apex, each petal with two small scales in the throat of the corolla.

Habitat and distribution
Hedgerows, waysides, almost always near habitation. Common in most areas.

Additional information
Perhaps native in the south-west where it grows along streams. Elsewhere probably introduced as an old cottage-garden herb, hence its association with habitation.

Pink family *Caryophyllaceae*
Flowering time: April–Oct.
Height of growth: 10–30 cm
Dicotyledonous; Perennial

Identification marks
The plant produces numerous prostrate or ascending stems which together form a dense turf or cushion. Flowers 1.5–2.5 cm in diameter, red. Calyx inflated-tubular, with short hairs. Leaves opposite.

Habitat and distribution
Not British. Rubble heaps, stony mats and rocky thickets. Indicator of lime. South-west and southern central Europe.

Additional information
The plant is occasionally planted in rockeries. This species is a popular rock-garden plant and very occasionally escapes.

Orpine
Sedum telephium

Water Avens
Geum rivale

Stonecrop family *Crassulaceae*
Flowering time: July–Sept.
Height of growth: 20–60 cm
Dicotyledonous; Perennial

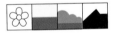

Identification marks
Leaves ovoid, smooth, flat but thick and fleshy. Stem erect. Inflorescence has many flowers packed together, often spread out in almost umbellate arrangement.

Habitat and distribution
In open woodland and hedgerows. Likes relatively dry stony ground rich in nutrients. Scattered throughout, in some parts local or rare; may also be found growing wild from gardens (old ornamental plant).

Rose family *Rosaceae*
Flowering time: May–Sept.
Height of growth: 20–60 cm
Dicotyledonous; Perennial

Identification marks
Several flowers in a lax cyme, drooping. The reddish-brown sepals are conspicuous. The petals are orange-red or yellowish. Leaves alternate, irregularly pinnate, upper ones trifoliate.

Habitat and distribution
Wet meadows, flat moors, ditches, light wet woodland. Common in northern England Wales and Scotland, local or rare in southern Scotland.

Additional information
The individual plants within the species vary very greatly, but so far it has not been possible to clearly define any further divisions. In larger groups one often finds 'anomalies': the bracts form an integral part of the flowers, the flower does not droop, etc

Marsh Cinquefoil
Potentilla palustris

Common Stork's-bill
Erodium cicutarium

Rose family *Rosaceae*
Flowering time: June–Sept.
Height of growth: 10–50 cm
Dicotyledonous; Perennial

Identification marks
Flower deep purple, almost brownish-red – not only the shorter narrower petals but also the sepals which are a dull purplish-red on the inside (but usually conspicuously green on the outside). Leaves palmately lobed with 5–7 lobes. Stem prostrate or erect.

Habitat and distribution
Fens, marshes and heaths. Usually found on acid soil which is flooded at least part of the year. Commoner in the north but widely distributed throughout Britain.

Additional information
Its classification in the genus *Potentilla* is sometimes disputed and it has been known as *Comarum palustre*. There are no known hybrids with other species of the genus.

Geranium family *Geraniaceae*
Flowering time: April–Oct.
Height of growth: 15–50 cm
Dicotyledonous; Perennial

Identification marks
Flowers approx. 1 cm in diameter, light purple to pink and often with lighter-coloured spots. Leaves pinnate, leaflets deeply dissected. Beak of the fruit 3–4 cm long with a ring-shaped constriction below the tip.

Habitat and distribution
Fields, sand dunes and waste land. Prefers a sandy soil. Scattered throughout Britain but commonest in coastal districts.

Additional information
The popular name refers to the beak-shaped fruit.

Cut-leaved Crane's-bill
Geranium dissectum

Long-stalked Crane's-bill
Geranium columbinum

Geranium family *Geraniaceae*
Flowering time: May–Oct.
Height of growth: 20–50 cm
Dicotyledonous; Annual – Biennial

Identification marks
Flowers in pairs, small, 8–10 mm in diameter, light purplish-red. Stem of inflorescence shorter than its bract. All leaves petiolate. Stem branched, ascending or erect. Leaves cleft almost to the base and composed of 5–7 lobes.

Habitat and distribution
Fields, gardens, wayside verges. Likes stony ground rich in nutrients, dry rather than moist. Scattered.

Additional information
Probably originates from the western Mediterranean area. From there it has been introduced and has established itself not only in Europe but also in the temperate zones of both the northern and the southern hemisphere.

Geranium family *Geraniaceae*
Flowering time: May–Sept.
Height of growth: 15–50 cm
Dicotyledonous; Annual

Identification marks
Flowers in pairs, small, 1.2–1.7 cm in diameter, light purple. Stems of inflorescence always longer than the bract. All leaves petiolate. Stem branched, erect or ascending. Leaves cleft almost to the base, 5–7 segments.

Habitat and distribution
Fields, gardens, path verges. Likes sandy rather dry, chalky soil. Scattered, but nowhere common, becoming rarer in the north.

Additional information
Probably originates from the eastern Mediterranean area. From there it has not only established itself in northern Europe but has spread across the entire northern hemisphere.

Hedgerow Crane's-bill
Geranium pyrenaicum

Herb Robert
Geranium robertianum

Geranium family *Geraniaceae*
Flowering time: June–Sept.
Height of growth: 15–50 cm
Dicotyledonous; Perennial

Identification marks
Flowers in pairs, small, 10–15 mm in diameter, purplish-violet or violet in colour. Stem of inflorescence longer than bract. Leaves have short petioles. Stem little branched, ascending or erect. Leaves roundish, kidney-shaped at the base, 3–7 cm in diameter, slightly indented to form 7–9 lobes.

Habitat and distribution
Field margins, pathways and hedgerows. Found in loamy soil. Scattered.

Additional information
Probably originates from the mountains around the Mediterranean. The plant has only been established in Britain for approximately 200 years.

Geranium family *Geraniaceae*
Flowering time: June–Oct.
Height of growth: 15–50 cm
Dicotyledonous; Annual – Biennial

Identification marks
Flowers in pairs, small, 1.4–1.8 cm in diameter, pink. Sometimes also very pale pink with darker veins. Stem of inflorescence much longer than the bract. Leaves cleft almost to the base in 3–5 leaflets, which are in turn divided almost to the central vein. Whole plant usually reddish tinged.

Habitat and distribution
Woodland, hedgerows, shingle beaches, walls, waste land. Common.

Additional information
In coastal areas there is a subspecies found on sand. It is smaller and has a prostrate stem (ssp. *maritimum*). *G. purpureum* is similar, lacks reddish tinge, has smaller flowers.

Marsh Crane's-bill
Geranium palustre

Bloody Crane's-bill
Geranium sanguineum

Geranium family *Geraniaceae*
Flowering time: June–Sept.
Height of growth: 20–100 cm
Dicotyledonous; Perennial

Identification marks
Flowers in pairs, flowers only 2.5–3 cm in diameter. Petals only slightly crenate, reddish-violet to pale purplish-red. No glandular hairs on the peduncles. Stem branched, ascending, with coarse hairs. Leaves have 7 palmate lobes.

Habitat and distribution
Not British. River banks, ditches, damp meadows and damp open woodland areas. Requires permanently moist soil.

Additional information
Marsh Crane's-bill is predominantly found in eastern Europe, as far east as Siberia and west as far as the Rhine.

Geranium family *Geraniaceae*
Flowering time: June–Aug.
Height of growth: 10–50 cm
Dicotyledonous; Perennial

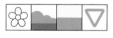

Identification marks
Flowers solitary, 2.5–3 cm in diameter, carmine. Petals crenate. Stem and peduncles have projecting hairs. Leaves 5–7 palmate lobes. Leaf lobes linear, 1 or 2 segments at the apex.

Habitat and distribution
Dry woodland, sand dunes, semi-dry turf. Likes sunny spots. Often found on limestone. Local but spread over most of Britain. Least common in the south. Usually in large numbers in localities where it occurs.

Wood Crane's-bill
Geranium sylvaticum

Dusky Crane's-bill
Geranium phaeum

Geranium family *Geraniaceae*
Flowering time: June–Sept.
Height of growth: 30–60 cm
Dicotyledonous; Perennial

Identification marks
Flowers in pairs, 2–3 cm in diameter, reddish-
or bluish-violet. Leaves 7–12 cm wide, usually
5–7 lobes with segments which do not go
right to the leaf base. Stem erect, usually
branched with a dense covering of hairs,
glandular above, bent downwards at the
bottom of the stem.

Habitat and distribution
Woodland, meadows and roadsides (up to
above 500 m in Scotland). Needs damp soil
rich in nutrients and humus. Never common
but widely distributed in suitable habitats.

Geranium family *Geraniaceae*
Flowering time: July–Oct.
Height of growth: 20–50 cm
Dicotyledonous; Perennial

Identification marks
Flowers in pairs, 2–3 cm in diameter,
brownish-violet in colour. In contrast to the
other European species the petals are spread
out flat or reflexed. Leaves 5–10 cm wide,
dissected into irregular lobes, the deepest cut
to 1/3 of the leaf width.

Habitat and distribution
Roadsides and hedge banks. Prefers moist
loamy soil low in lime. Rare.

Additional information
Native of southern Europe. North of the Alps
it is only found growing wild as a garden
escape but has established itself locally in
England, Wales and southern Scotland.

Musk Mallow
Malva moschata

Common Mallow
Malva sylvestris

Mallow family *Malvaceae*
Flowering time: July–Sept.
Height of growth: 30–60 cm
Dicotyledonous; Perennial

Identification marks
Flowers pale red, pink or almost white, 4–5 cm in diameter. Outer sepals 3–5 mm long and only about 1 mm wide. Upper leaves palmately lobed into 5–7 segments almost to the base of the leaf. Hairs on the stem unbranched (magnifying glass needed).

Habitat and distribution
Dry grassland, hedgebanks. Requires soil rich in nutrients. Not uncommon, occurring throughout Britain though never in large numbers.

Additional information
A variable plant especially in the way the leaves are dissected. Similar: *M. alcea*, outer sepals oval, 4–8 mm long and 2–4 mm wide. Rarely found and not persisting.

Mallow family *Malvaceae*
Flowering time: July–Sept.
Height of growth: 20–120 cm
Dicotyledonous; Perennial

Identification marks
2–6 flowers in a raceme borne in the leaf axils, reddish-violet in colour, approx. 4 cm in diameter. Petals vary slightly in size, and have a notched margin. Upper leaves palmately dissected but not for more than 2/3 of their length. Stem erect. Leaf lobes crenate.

Habitat and distribution
Roadsides, hedgerows and waste places. Nitrogen indicator. Widely distributed and common in southern England, rarer further north.

Additional information
Native of the Mediterranean region but has been known in Britain since Roman times and is usually considered native though its exact status is doubtful.

Dwarf Mallow
Malva neglecta

Wild Angelica
Angelica sylvestris

Mallow family *Malvaceae*
Flowering time: June–Sept.
Height of growth: 30–50 cm
Dicotyledonous; Biennial – Perennial

Identification marks
Flowers in axillary clusters, pale pinkish-red to almost white. Flower does not open wide but is funnel-shaped. The flower diameter is therefore not relevant, only the length of the flower. This is between 8–15 mm. The flower is considerably longer than the calyx especially on those plants which remain small. The flower stalk is bent over. Upper leaves are palmately lobed but not for more than 2/3 of their length. Stem usually prostrate.

Habitat and distribution
Weedy areas along roadside and in waste places. Widespread in southern England but never common.

Additional information
Similar: Small-flowered Mallow (*M. pusilla*), with flowers only about 5 mm long and never longer than the sepals. Occasionally found in waste places.

Umbellifer family *Umbelliferae*
Flowering time: July–Sept.
Height of growth: 30–200 cm
Dicotyledonous; Perennial

Identification marks
Flowers in double umbels, each with 20–40 rays, either without bracts or reduced to only a few and falling early. Secondary umbels with numerous bracteoles. Stem hollow, glaucous-blue to reddish, tinged with white. Leaves 2–3 pinnate, petiolate and with conspicuously inflated leaf sheaths.

Habitat and distribution
Frequent in fens and marshes along the banks of rivers and ditches, in wet meadows as well as in wet woodlands.

Additional information
Wild Angelica was a popular medicinal plant and is one of the many *Umbelliferae* which have either white or reddish flowers (p. 69; p. 368).

Bird's-eye Primrose
Primula farinosa

Primula clusiana

Primrose family *Primulaceae*
Flowering time: May–July
Height of growth: 5–20 cm
Dicotyledonous; Perennial

Primrose family: *Primulaceae*
Flowering time: May–July
Height of growth: 2–5 cm
Dicotyledonous; Perennial

Identification marks
Rose-purple flowers, carried in umbels, approx. 1 cm in diameter. The 5 petals are spread out flat; flower stalks are mealy when young. The calyx is small. The leaf margins are crenulate and the undersides are covered with yellowish or white meal.

Habitat and distribution
Needs peaty soil in calcareous areas. Found from Derbyshire to Scotland where it is sometimes abundant.

Additional information
Similar: Long-flowered Primrose (*P. halleri*), corolla tube usually 2–3 cm long (makes it easy to identify), calyx 1–1.5 cm long. Not known in Britain. South-eastern Alps, western Alps. Very rare.

Identification marks
Usually only 2 flowers on each peduncle. 3 cm or more in diameter. Flower lobes not quite flat but partly bell-like in shape. They are divided to 1/3–1/2 of their length. Leaves slightly sticky at most, light green above and grey-green below. The margin is frequently lighter in colour with a narrow scarious edge.

Habitat and distribution
Not British. Needs stony ground rich in lime which is covered with snow for long periods.

Scarlet Pimpernel
Anagallis arvensis

Sea Milkwort
Glaux maritima

Primrose family *Primulaceae*
Flowering time: July–Oct.
Height of growth: 8–15 cm
Dicotyledonous; Annual

Identification marks
Flowers 5–7 mm in diameter, usually brick red. Petal margins usually slightly crenate. Flowers borne on slender stems from the leaf axils. Stem generally prostrate. Leaves opposite, occasionally in whorls of 3, sessile, ovate.

Habitat and distribution
Root crop fields, gardens, waste places. Indicator of light soils. Needs good supply of nutrients. Common.

Additional information
There is also a blue subspecies which should not be confused with *A. foemina*. The petals of all forms of the Scarlet Pimpernel are only slightly crenate.

Primrose family *Primulaceae*
Flowering time: May–June
Height of growth: 5–15 cm
Dicotyledonous; Perennial

Identification marks
The flowers are stemless and solitary in the axils of the leaves. They reach a diameter of approx. 8 mm. Usually pale pink, somewhat darker towards the throat of the corolla. The leaves grow to 5 mm long and 3 mm wide. They are extremely fleshy and grow very close to the short stems. These are usually prostrate, spreading and rooting.

Habitat and distribution
Needs salty, sandy or silty ground. A seashore plant found mostly at the edge of salt marshes or cliffs. Frequent around Britain.

Additional information
Sea Milkwort grows along virtually all coasts in the northern hemisphere, may be rarer in some locations, i.e. the Baltic coast.

Common Cyclamen
Cyclamen purpurascens

Sea Pink
Armeria maritima

Primrose family *Primulaceae*
Flowering time: June–Sept.
Height of growth: 5–20 cm
Dicotyledonous; Perennial

Identification marks
Flowers solitary on their peduncles, their recurved petals very conspicuous. The long peduncles curl up like corkscrews while the fruit is forming. Strong scent. Leaves evergreen, reni- or cordiform. Leaf margin slightly crenate.

Habitat and distribution
Not British. Needs somewhat damp soil containing humus and rich in nutrients. Forest in the Alpine foreland and in the Alps. Very rare.

Additional information
Occasionally Cyclamen may initially be confused with *Erythronium dens-canis* which has 6 recurved petals. Not British.

Sea Lavender family *Plumbaginaceae*
Flowering time: May–Sept.
Height of growth: 20–40 cm
Dicotyledonous; Perennial

Identification marks
Inflorescence 1.5–2.5 cm wide, borne in somewhat flattened compact globular heads, containing 10–30 flowers, usually pink but may also be purplish-red. Petal lobes crenate. Leaves approx. 1–2 mm wide and 5–10 cm long and are thick in texture.

Habitat and distribution
Found on salt marshes and shingle banks. Tolerates high salt content. Common on cliffs and widely distributed around British coastline. Also found in peaty and rocky areas of some Scottish mountains to 1,400 m.

Common Centaury
Centaurium erythraea

Lesser Bindweed
Convolvulus arvensis

Gentian family *Gentianaceae*
Flowering time: July–Oct.
Height of growth: 10–50 cm
Dicotyledonous; Annual

Identification marks
10–40 flowers in lax, umbel-like inflorescence. Flowers pinkish-red with slight tinge of purplish-violet, approx. 1 cm across. Stem erect, always simple, branched only around inflorescence. Basal leaves in a rosette. Stem leaves elongate-ovate.

Habitat and distribution
Dunes, woodland margins, dry turf. Common, especially in England.

Additional information
Similar: *C. pulchellum*. Stem usually branched right from the base. No basal rosette leaves. Smaller (3–15 cm). Likes damp, even wet habitat in marshy meadows. Local, mostly near the sea in southern England and Ireland.

Convolvulus family *Convolvulaceae*
Flowering time: June–Oct.
Height of growth: 30–100 cm
Dicotyledonous; Perennial

Identification marks
Flowers usually solitary, rarely 2–3 in leaf axils; funnel-shaped, slight scent. Stem prostrate or twining to the left, glabrous. Leaves alternate, petiolate, sagittate or hastate at the base.

Habitat and distribution
Fields, gardens, pathways, waste ground. Nitrogen indicator. Common.

Additional information
Roots, up to 2 m long, used to store nutrients and then provide for further growth if stems and leaves are destroyed.

Dodder
Cuscuta europaea

Common Comfrey
Symphytum officinale

Dodder family *Cuscutaceae*
Flowering time: June–Sept.
Height of growth: 20–100 cm
Dicotyledonous; Annual

Identification marks
The host plant (Nettle, Hedge Bindweed, Mugwort) is surrounded and penetrated by a network of thin thread-like stems. Flowers, pale red and sometimes almost white, in tangled clusters. Often only seen when the plant is closely inspected.
Habitat and distribution
Wasteland, river banks. Rare.
Additional information
All species parasites. Similar: Common Dodder (*C. epithymum*), host plants Thyme, Broom, Heather; sometimes found on heaths. *C. pilinum*, on flax. Only Common Dodder is likely to be seen in Britain.

Borage family *Boraginaceae*
Flowering time: May–June
Height of growth: 30–120 cm
Dicotyledonous; Perennial

Identification marks
Entire plant covered with rough hairs. Leaves narrowly ovoid, distinctly decurrent. Flowers small, campanulate, drooping, in scorpioidal cymes.
Habitat and distribution
On damp to wet ground, always rich in nutrients. In wet meadows, on river banks and in ditches. Found throughout Britain but less so in the north.

Pulmonaria obscura

Vervain
Verbena officinalis

Borage family *Boraginaceae*
Flowering time: March–April
Height of growth: 15–40 cm
Dicotyledonous; Perennial

Identification marks
Several flowers together in umbellate-like head. Peduncles of axillary inflorescences are as long as their bracts. Flowers like cowslips, opening red, becoming violet and finally blue as they fade. No narrowing of basal leaves. No spots on the leaves.

Habitat and distribution
Not British. Woodland. Likes calcareous loamy soil. Scattered and local in northern Europe. Usually abundant where it occurs.

Additional information
Similar: Lungwort (*P. officinalis*), peduncles of axillary inflorescence shorter than the respective bract. Spotted leaves. Not British. Alps, Alpine foreland. Rare.

Verbena family *Verbenaceae*
Flowering time: July–Oct.
Height of growth: 30–60 cm
Dicotyledonous; Annual – Perennial

Identification marks
Numerous small (3–5 mm long) reddish-violet or pale lilac flowers in a spike. Inflorescence conspicuously squarrosely branched. Leaves deeply divided, upper ones less so.

Habitat and distribution
Roadsides and waste places. Nitrogen indicator. Local. In Britain commonest in the south.

Additional information
Vervain is thought to have originated from the Mediterranean area. Requires warmth.

Deadly Nightshade
Atropa bella-donna

Mountain Valerian
Valeriana montana

Nightshade family *Solanaceae*
Flowering time: June–July
Height of growth: 5–150 cm
Dicotyledonous; Perennial

Identification marks
Single flowers in axils of upper leaves. They have a greenish-red tinge, lobes are deep brownish-red, brownish-violet or purplish-violet. Stem erect, leaves ovate, decurrent. Usually a large leaf grows next to a small leaf. Fruit a berry, large and black.

Habitat and distribution
Open woodland areas especially on chalky soils. Needs rather moist soil rich in nutrients. Rather rare.

Additional information
Deadly Nightshade is deadly poisonous. It contains hyoscyamine and smaller quantities of atropine.

Valerian family *Valerianaceae*
Flowering time: April–July
Height of growth: 10–60 cm
Dicotyledonous; Perennial

Identification marks
Entire inflorescence almost flat, umbellate with many individual flowers. Individual flowers up to 2.5 mm long, usually pale pink, rarely entirely white. Stem erect, hollow. Leaves on the non-flowering shoots narrow abruptly towards the stem. Only solitary leaves have 3 segments, normally they are ovate, lanceolate with 3–8 leaf pairs on one stem.

Habitat and distribution
Not British. Stony places or scree; Alps, Alpine foreland. Scattered.

Additional information
Similar: Three-leaved Valerian (*V. tripteris*). Stem usually contains floury substance, not hollow. Scree and stony places. Not British. Alps, Alpine foreland and higher upland areas in southern central Europe. Rare.

Marsh Valerian
Valeriana dioica

Common Valerian
Valeriana officinalis

Valerian family *Valerianaceae*
Flowering time: May–July
Height of growth: 10–30 cm
Dicotyledonous; Annual

Identification marks
Flowers in small clusters in rounded head at the end of the stem. Flowers dioecious: male and female flowers are found on separate plants. Male larger than the female, flowers pinkish. Basal leaves entire, stem leaves pinnately divided.

Habitat and distribution
Banks of rivers, damp meadows, fens and boggy areas. Scattered throughout Britain in suitable habitats.

Additional information
Similar: Tuberous Valerian (*V. tuberosa*). Flowers usually bisexual. Basal leaves at least 1½ times as long as wide, usually 3 times longer than wide. Rhizome forms tubers. Only the southern Alps. Rare.

Valerian family *Valerianaceae*
Flowering time: July–Sept.
Height of growth: 30–170 cm
Dicotyledonous; Perennial

Identification marks
Flowers in terminal whorled panicle of considerable size. Flowers bisexual, flesh pink to very pale whitish-red, scented. All leaves opposite (6–9 pairs), pinnate (15–21 leaflets), glabrous.

Habitat and distribution
Damp areas in woodland, in meadows and ditches. Also found on dry calcareous soils. Widely distributed though nowhere common.

Additional information
Similar: *V. pratensis*: stem has only 5–7 leave pairs; leaves usually short with solitary hairs. Stem glabrous. Moors, meadows, marshy woodland. Likes the warmth. Very rare; not British.

Flowering Rush
Butomus umbellatus

Meadow Saffron
Colchicum autumnale

Flowering Rush family *Butomaceae*
Flowering time: June–Aug.
Height of growth: 50–150 cm
Monocotyledonous; Perennial

Identification marks
10–30 flowers in a lax umbellate inflorescence, up to 2–2.5 cm across. Outer petals shorter and narrower than inner ones. Inner petals usually have conspicuous dark red veins. Basal leaves reed-like, fluted, stiffly erect.

Habitat and distribution
Reed beds in stagnant or slow-flowing water, mud ditches, canals and river margins. Local but scattered throughout Britain, except Scotland. Likes the warmth.

Additional information
Flowering Rush can be found almost everywhere but is not common anywhere; despite its extensive area of distribution virtually no variations within the species are known.

Lily family *Liliaceae*
Flowering time: Aug.–Oct.
Height of growth: 10–50 cm
Monocotyledonous; Perennial

Identification marks
Flowers on whitish 'stalk'. Free section of the petals 4–7 cm long. No leaves present during flowering time. Leaves, like tulip leaves appear in the following spring.

Habitat and distribution
Meadows, damp open woodland. Likes nitrogen. Rare but in abundance in a few places in England.

Additional information
Similar: *C. alpinum*, free section of the petal only 2–3 cm long. *C. bulbocodium*, flower appear in the spring together with leaves which are hood-shaped at the apex. Very rare Not British. All species are very poisonous.

Crow Garlic
Allium vineale

Martagon Lily
Lilium martagon

Lily family *Liliaceae*
Flowering time: June–Aug.
Height of growth: 30–60 cm
Monocotyledonous; Perennial

Identification marks
Inflorescence a false umbel usually with bulbils instead of flowers. Where flowers form these have long peduncles and the leaves are like chives. Stem is covered with hairs to about half-way up.
Habitat and distribution
Roadsides, fields. Can be a serious weed in arable land. Common in England and Wales.
Additional information
Similar: Round-headed Leek (*A. sphaerocephalum*). Inflorescence rarely has bulbils, only flowers with very short peduncles. Dry turf. Rare. Only known in 2 sites in Britain.

Lily family *Liliaceae*
Flowering time: June–Aug.
Height of growth: 30–120 cm
Monocotyledonous; Perennial

Identification marks
2–10 flowers, very rarely more, in a lax raceme. The flowers droop; 5–7 cm across. Petals have dark brown-red spots and when in full bloom petals are folded back. Leaves narrow ovoid, margin entire, in whorls.
Habitat and distribution
Open woodland. Likes calcareous soil containing nutrients. Not native to Britain but naturalized in a few places.
Additional information
It is native to much of Europe eastwards to Siberia and Mongolia.

Pheasant's Eye
Adonis aestivalis

Houseleek
Sempervivum tectorum

Buttercup family *Ranunculaceae*
Flowering time: May–July.
Height of growth: 30–50 cm
Dicotyledonous; Perennial

Identification marks
Flowers solitary, scarlet, more rarely yellowish-red. Petals spread out. Sepals clinging. Small fruit without black spots. Leaves bi- or tripinnate.
Habitat and distribution
Fields, waste ground. Likes dry calcareous soil. In Britain it has never been more than of casual occurrence.
Additional information
Similar: *A. autumnalis* (*A. annua*), sepals horizontal or projecting backwards, flowers usually dark red. Garden plant originating from the Mediterranean region, occasionally a weed in cornfield. Now rarely seen.

Stonecrop family *Crassulaceae*
Flowering time: July–Sept.
Height of growth: 10–60 cm
Dicotyledonous; Perennial

Identification marks
The peduncle of the inflorescence rises from a dense rosette of basal leaves and is covered with scale-like leaves. It bears 10–25 flowers which are all closely compressed; 1–2 cm across, 10–20 petals.
Habitat and distribution
Rocks and walls. Only grows wild in the Alps. Otherwise an ornamental plant which has become naturalized. Very rare.

Purple Loosestrife
Lythrum salicaria

Dwarf Snowbell
Soldanella pusilla

Loosestrife family *Lythraceae*
Flowering time: July–Sept.
Height of growth: 60–160 cm
Dicotyledonous; Perennial

Identification marks
Flowers in whorls on a long spike; 1.2–1.8 cm across. Leaves opposite, decussate, occasionally 3 in a whorl, rounded at the base. Stem erect, quadrangular, also covered with short hairs like the leaves.

Habitat and distribution
River banks, reed beds, fens, wet meadows, wet woodland. Common.

Additional information
Similar: Slender Loosestrife (*L. virgatum*), leaves not rounded, narrower at the base. Entire plant glabrous. Leaves long, slender and grow closer to the stem. Not British. Eastern Europe to Austria. Rare.

Primrose family *Primulaceae*
Flowering time: May–Aug.
Height of growth: 2–10 cm
Dicotyledonous; Perennial

Identification marks
A single flower on each peduncle. Constantly drooping; pale pinkish-violet. Campanulate flower (approx. 1–1.5 cm long) is cut into a fringe of numerous segments to about 1/4 of its length. Leaves are round to reniform, never wider than 1 cm, margin entire, never thick.

Habitat and distribution
Not British. Often under late snow patches in the central mountain ranges, especially east of the Rhine in the Alps, scattered.

Additional information
The Snowbell species resemble each other. *S. minima* nearly always has numerous small glandular hairs on the peduncle. Not British.

Hemp Agrimony
Eupatorium cannabinum

Adenostyles alliariae

Daisy family *Asteraceae (Compositae)*
Flowering time: July–Sept.
Height of growth: 40–150 cm
Dicotyledonous; Annual – Perennial

Identification marks
Flowers in dense corymbs at the top of long stems. All flowers tubular in shape. Some leaves opposite having 3–5 segments. Leaflets lanceolate, coarsely serrate.

Habitat and distribution
Moist woodland areas, banks of rivers, fens and marshes. Moisture indicator, somewhat calcicolous. Common, growing in colonies.

Additional information
Previously used as a medicinal plant. Its medicinal value is, however, disputed, although the plant does contain an unidentified bitter substance.

Daisy family *Asteraceae (Compositae)*
Flowering time: June–Aug.
Height of growth: 50–150 cm
Dicotyledonous; Perennial

Identification marks
Flower heads in umbel-like axillary inflorescences; these grow so closely together that they virtually form an overall umbrella-like inflorescence. The individual heads contain only 3–6 flowers. The uppermost stem leaf is usually sessile with diffuse base. Leaf undersides have a felt-like covering of hairs which can be rubbed off.

Habitat and distribution
Not native to Britain. Alps, occasionally also in the Black Forest and in the Vosges in open woodland and in high bushy areas. Scattered

Additional information
Similar: *A. glabra*, all leaves petiolate. Capitula usually have 3 flowers. Alps. Scattered or limestone. Not native to Britain.

Butterbur
Petasites hybridus

Purple Coltsfoot
Homogyne alpina

Daisy family *Asteraceae (Compositae)*
Flowering time: March–April
Height of growth: 30–60 cm
Dicotyledonous; Annual

Identification marks
Flowers in racemose capitula. Flowers appear before the leaves. They are dioecious. Only leaf scales on peduncle. The very large leaves appear after flowering period. They are round and 60 cm across.

Habitat and distribution
River banks, ditches, river meadows, forest margins. The male plant is common through the British Isles, the female limited to parts of northern England.

Additional information
Similar: Alpine Butterbur (*P. paradoxus*), leaves appear at the end of the flowering period but only grow to 20 cm across. Undersides constantly covered with white felt-like hairs. Conspicuously triangular in shape. Not British. Limestone scree slopes in the Alps. Rare.

Daisy family *Asteraceae (Compositae)*
Flowering time: May–Aug.
Height of growth: 10–30 cm
Dicotyledonous; Perennial

Identification marks
Stems carry a single flower 2–3 cm across. The thread-like flowers are purplish-red. Basal leaves are round-reniform, petiolate, toothed, without any felty covering of hairs. Scale-like stem leaves found on the lower third of the stem. Plant often has runners.

Habitat and distribution
Rare on 2 mountain ledges in Scotland. Usually considered an introduction.

Additional information
Similar: *H. discolor*, capitula, around 1 cm across, basal leaves covered with thick felt. Scale-like stem leaves found in middle third of stem. No runners. Not British. Limestone Alps. Dolomites. Scattered.

Mountain Everlasting
Antennaria dioica

Yarrow
Achillea millefolium

Daisy family *Asteraceae (Compositae)*
Flowering time: May–June
Height of growth: 8–25 cm
Dicotyledonous; Perennial

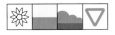

Identification marks
3–12 small capitula in terminal umbel-like racemes which contain disc florets. Plant is dioecious. Bracts of the capitula containing male flowers are white, those containing female flowers pink. Basal leaves in a rosette; runners above the ground.

Habitat and distribution
Dry heaths and mountain slopes often in leached soil over limestone. Frequent in northern Britain.

Additional information
Similar: Carpathian Cat's-foot (*A. carpatica*), bracts of the male capitula white only at the tips and along the margin; on the female flowers they are translucent, brownish. No runners. Not British. Very stony Alpine turf. Rare.

Daisy family *Asteraceae (Compositae)*
Flowering time: June–Aug.
Height of growth: 8–45 cm
Dicotyledonous; Perennial

Identification marks
Stem erect, leaves alternate, bi- or tripinnate, leaflets divided into 2–5 parts. Flowers in capitula, arranged in loose corymbs: on the inside yellowish-white disc florets, on the outside usually only 4–5 white or rarely red ray florets.

Habitat and distribution
Common in meadows, semi-dry turf, banks, pastures, arable fields and along waysides and hedgerows. Prefers loose loamy soils which are rich in nutrients and not too moist.

Additional information
The colour of the ray florets varies from white through pure white to reddish pink and deep red.

Alpine Thistle
Carduus defloratus

Musk Thistle
Carduus nutans

Daisy family *Asteraceae (Compositae)*
Flowering time: June–Oct.
Height of growth: 30–120 cm
Dicotyledonous; Perennial

Identification marks
Stem unbranched or only with individual branches. Each stem bears 1 erect head, 1–3 cm across, containing red disc florets. Leaves pinnate, with soft thorns. Upper part of stem has no foliage or only small leaves.

Habitat and distribution
Not British. Open woodland, thickets, mountain meadows, moderately hilly areas in southern Germany, Alps. Rare.

Daisy family *Asteraceae (Compositae)*
Flowering time: July–Aug.
Height of growth: 30–100 cm
Dicotyledonous; Perennial

Identification marks
Little branched stem; at the end of each branch a single drooping head 3.5–7 cm across. It contains red disc florets only. Leaves pinnate, with long thorns each at least 4 mm long, sometimes even 6 mm.

Habitat and distribution
Fields, waysides, waste ground, usually on chalky soils. Scattered throughout but frequent only in the south.

Additional information
Similar: *C. platylepis*, drooping flower heads only 3–4 cm across. Leaves pinnate, thorns less than 3 mm long. Not British. Paths, Alpine pastures. Only southern and western Alps. Rare.

Spear Thistle
Cirsium vulgare

Marsh Thistle
Cirsium palustre

Daisy family *Asteraceae (Compositae)*
Flowering time: June–Oct.
Height of growth: 60–130 cm
Dicotyledonous; Biennial

Identification marks
Stem little branched; at the end of each branch is a single head 2–4 cm across, containing red disc florets. The bract beneath each head has a light cobweb-like covering of hairs. Leaves are conspicuously decurrent, with a fine web-like covering on the undersides.
Habitat and distribution
Roadsides, waste places. Likes calcareous soil rich in nitrogen. Common.
Additional information
A variety of the Spear Thistle (*C. hypoleucum*) has the undersides of its leaves covered with white felt.

Daisy family *Asteraceae (Compositae)*
Flowering time: July–Sept.
Height of growth: 90–200 cm
Dicotyledonous; Biennial

Identification marks
Stem little branched; at the end of each branch there are several sessile or short stalked flower heads, about 1 cm across and 1.5 cm long. Leaves markedly decurrent; stem looks as though it has prickly wings. Leaves also have very long prickles.
Habitat and distribution
Marshes, wet meadows and damp woodlands. Likes moisture and avoids very calcareous soil. Common.
Additional information
The Marsh Thistle can vary greatly, depending on the habitat, particularly in height, leaf shape and the number of prickles.

Stemless Thistle
Cirsium acaule

Saw-wort
Serratula tinctoria

Daisy family *Asteraceae (Compositae)*
Flowering time: July–Sept.
Height of growth: 5–20 cm
Dicotyledonous; Perennial

Identification marks
Inflorescence single, normally stalkless head, 2.5 cm long and almost equally wide. It contains only red disc florets. Leaves pinnatifid and spine-tipped, the lobes ovoid.

Habitat and distribution
Grasslands, particularly on chalk and limestone. Locally common in southern England.

Additional information
C. caulescens has flowers with stems to 20 cm high. The plant is found throughout southern and western Europe. Rare.

Daisy family *Asteraceae (Compositae)*
Flowering time: July–Sept.
Height of growth: 30–100 cm
Dicotyledonous; Perennial

Identification marks
Stem with few branches; flowers in a lax corymb of 10–40 small flowers. The stalked flowers barely reach 1 cm across but grow to a length of 1.5 cm. Purplish-red disc florets sometimes tend towards a violet colour. The leaves are undivided or pinnate and conspicuously sharply serrate.

Habitat and distribution
Rather moist open woodland, and grassland especially on chalky soils. Local but never common throughout England and Wales. Rare in Scotland.

Common Knapweed
Centaurea nigra

Brown Knapweed
Centaurea jacea

Daisy family *Asteraceae (Compositae)*
Flowering time: July–Sept.
Height of growth: 20–80 cm
Dicotyledonous; Perennial

Identification marks
Stem stout with few branches. Each branch bears one head which may reach 2–4 cm across. There are no enlarged florets around the margin of the flower. Below the flower head the stem is sometimes thickened. Leaves oval to lanceolate. Stem usually erect.

Habitat and distribution
Grassland, roadsides and waste places.

Additional information
Within the species several subspecies are distinguished: ssp. *nemoralis*, height over 60 cm, relatively free branching with little swelling beneath the flowers. All the bract appendages on ssp. *nigra* are black.

Daisy family *Asteraceae (Compositae)*
Flowering time: June–Oct.
Height of growth: 30–100 cm
Dicotyledonous; Perennial

Identification marks
Stem only branched at the top and not very freely. Branches with one head at the end up to 3–5 cm across. Head contains both disc florets and marginal florets which are conspicuously larger. Middle and lower leaves alternate, not divided. Lower leaves often notched to pinnately divided.

Habitat and distribution
Grassland and waste places. Not native to Britain but established in a few places in southern England.

Greater Knapweed
Centaurea scabiosa

Downy Burdock
Arctium tomentosum

Daisy family Asteraceae (Compositae)
Flowering time: July–Aug.
Height of growth: 30–130 cm
Dicotyledonous; Perennial

Identification marks
Stem little branched. Single flower head at the end of each branch up to 3.5–6 cm across. The heads contain disc and marginal florets, the latter being conspicuously the larger. Bracts have a dark brown edge. All stem leaves pinnate.
Habitat and distribution
Dry turf, meadows, cliffs, roadsides especially on chalky soils. Common.
Additional information
Commonest in the south but occurs throughout Britain. In Europe it is found from Finland to the Caucasus and in western Asia.

Daisy family Asteraceae (Compositae)
Flowering time: July–Aug.
Height of growth: 50–150 cm
Dicotyledonous; Biennial

Identification marks
5–50 flowers in a dense corymb at the end of the stem. Heads, covered with thick layer of web-like hairs, are up to 2–3 cm across. They contain only disc florets. The bracts are hooked (burs). Leaves very large, rounded or heart-shaped at the base.
Habitat and distribution
Roadsides and waste places especially the banks of rivers. Likes calcareous loamy soil rich in nitrogen. Rare, casual.

Arctium nemorosum

Lesser Burdock
Arctium minus

Daisy family *Asteraceae (Compositae)*
Flowering time: July–Aug.
Height of growth: 100–250 cm
Dicotyledonous; Biennial

Identification marks
3–5 heads, 3–4.5 cm across, a cluster at the end of the stems. Stems may be covered by very fine layer of web-like hairs, usually glabrous. Bracts only slightly hooked. Leaves very large, those at the base slightly heart-shaped.

Habitat and distribution
Woodland, usually in open areas or in clearings, roadside and waste places. Scattered.

Additional information
Easy to identify in typical habitats as most climbing species do not occur. Nevertheless hybrids have been recorded, mainly near to residential areas located in woodland.

Daisy family *Asteraceae (Compositae)*
Flowering time: July–Sept.
Height of growth: 50–130 cm
Dicotyledonous; Biennial

Identification marks
5–12 heads in a cluster at stem end; with light covering of web-like hairs, 1–2.5 cm across. They contain disc florets only. Bracts hooked. Branches erect and spreading. Stems leafy, the leaves being large and rounded or slightly heart-shaped at the base.

Habitat and distribution
Paths, waste places. Likes loamy soil. Nitrogen indicator. Scattered.

Additional information
Similar: Greater Burdock (*A. lappa*): 5–12 heads in an umbel-like arrangement on the stem. Each head grows to 3–4 cm across. Pathways and waste places. Nitrogen indicator. Scattered.

Red Hare's Lettuce
Prenanthes purpurea

Orange Hawkweed
Hieracium aurantiacum

Daisy family *Cichoriaceae (Compositae)*
Flowering time: July–Aug.
Height of growth: 60–160 cm
Dicotyledonous; Biennial

Identification marks
Numerous heads in a very lax panicle, each containing only 3–5 ray florets. The elegant heads 1.5–2 cm across. Stem erect, branching only at the very top. Leaves glabrous with heart-shaped amplexicaul base, undersides bluish-green. Lower leaves elongate-lanceolate, notched. Margin of upper leaves is usually entire.

Habitat and distribution
Not British. Native to mainland Europe in woodland. Likes soil rich in humus but not containing too much lime.

Additional information
Those varieties where the leaf margin is always entire and the leaf narrow are classed as ssp. *angustifolia*, to distinguish them from the 'typical' forms.

Daisy family *Cichoriaceae (Compositae)*
Flowering time: June–Aug.
Height of growth: 20–50 cm
Dicotyledonous; Perennial

Identification marks
5–20 heads in a panicle on a branched stem. Heads have glandular hairs. Ray florets orange-yellow to brownish red, usually deep purplish-red when dried. Leaves in rosette arrangement, elongate and tongue-shaped, margin entire or slightly toothed, hairy.

Habitat and distribution
Not British. Alpine mats and thickets. Frequent ornamental plant; away from the Alps growing wild on poor meadows and along paths. In the Alps scattered, otherwise rare.

Additional information
Similar: *H. caespitosum*, usually 15–30 capitula. Flowers only deep yellow, not clearly orange-yellow. Do not become purplish-red when dried. Not British. Scattered.

Marsh Gladiolus
Gladiolus palustris

Red Helleborine
Cephalanthera rubra

Iris family *Iridaceae*
Flowering time: June–July
Height of growth: 30–70 cm
Monocotyledonous; Perennial

Identification marks
3–8 flowers in an unbranched spike. They grow to 2–3 cm long, are purplish-red and usually turn bluish when dried. Flowers incline downwards. Stem erect. Leaves 0.5–1 cm wide.

Habitat and distribution
Moorland. Dislikes fertilizer. Very rare in northern Europe.

Additional information
Similar: *G. imbricatus*, 3–8 flowers in a lax unilateral spike. Flowers grow to 2 cm long. Leaves 1–2 cm wide. Moorland, thickets. Occasionally cultivated in Britain, but not native.

Orchid family *Orchidaceae*
Flowering time: May–July
Height of growth: 20–80 cm
Monocotyledonous; Perennial

Identification marks
4–12 flowers in a lax spike. They are pinkish-red or purplish-red, occasionally with a tinge of violet. The lip does not have a spur. The upper leaves are erect, the lateral leaves are spread far apart when the flowers are in full bloom. Leaves ovoid-elongate, 6–12 cm long and 2–4 cm wide.

Habitat and distribution
Woodland. Likes warmth. Needs well drained soil containing mull. Rare.

Additional information
Red Helleborine is undoubtedly one of the most beautiful orchids native to Britain.

Dark Red Helleborine
Epipactis atrorubens

Pyramidal Orchid
Anacamptis pyramidalis

Orchid family *Orchidaceae*
Flowering time: June–Aug.
Height of growth: 30–60 cm
Monocotyledonous; Perennial

Identification marks
10–30 flowers in a lax somewhat 1-sided spike. Flowers approx. 1 cm across, purplish-red with a tinge of brown or violet. Lip does not have spur. A lateral constriction divides it into a front and a rear section, giving it a 'folded' look. Outer petals form a wide bell. Leaves elongate-ovoid.
Habitat and distribution
Woodland, rocks and screes. Likes calcareous soils. Rare.
Additional information
Similar: *E. purpurata*, 30–60 flowers in a compact cluster. Flowers approx. 1.5 cm across or wider, whitish to reddish in colour. Entire plant tinged with violet. Woodland. Rare.

Orchid family *Orchidaceae*
Flowering time: June–July
Height of growth: 20–50 cm
Monocotyledonous; Perennial

Identification marks
Inflorescence initially decidedly pyramidal, then somewhat cylindrical (when the lower flowers have faded). Flowers have thread-like spur equalling the ovary in length. Flowers deep rose-purple in colour. Lip is paler, and wider than it is long, divided into 3. Leaves narrow-lanceolate.
Habitat and distribution
Grassland on chalk and limestone; dry open woodland and thickets, more rarely on stable dunes. Locally frequent.
Additional information
When the Pyramidal Orchid occurs together with the Fragrant Orchid or with members of the *Orchis* species hybrids may well be found.

Round-headed Orchid
Traunsteinera globosa

Orchid family *Orchidaceae*
Flowering time: June–July
Height of growth: 20–50 cm
Monocotyledonous; Perennial

Identification marks
Inflorescence markedly globular or oval. It contains 80–120 flowers which barely reach a length of 1 cm. Flowers pale pink or purplish-red, often with a tendency towards violet. Lip elongate to trilobed, with dark spots. Spur slender. Outer petals drawn out into a conspicuously slender tip which is thickened at the front in the shape of a club.
Habitat and distribution
Not British. Mountains and outskirts of relatively hilly areas. Grows on meadows or semi-dry turf which is rich in nutrients but unfertilized. Rare.
Additional information
Native to Poland southwards to Spain, north-central Italy and southern Bulgaria.

Green-winged Orchid
Orchis morio

Orchid family *Orchidaceae*
Flowering time: May–June
Height of growth: 8–40 cm
Monocotyledonous; Perennial

Identification marks
Inflorescence a lax spike containing 4–12 flowers. These are relatively large. Lip alone up to 1 cm or more. It is wider than it is long and divided into four lobes. Spur short, fat and projects horizontally. Remaining petals bend inwards to form a conspicuous helmet shape. Stem erect, leaves elongate-lanceolate.
Habitat and distribution
Meadows and pastures, especially on limestone soils. Locally abundant, rarer in the north.
Additional information
First orchid to flower in the year. Has decreased over the last few decades owing to a lack of suitable habitats.

Burnt Orchid
Orchis ustulata

Military Orchid
Orchis militaris

Orchid family *Orchidaceae*
Flowering time: May–June
Height of growth: 8–20 cm
Monocotyledonous; Perennial

Identification marks
Dense globular to conical flower spike. Blossoms very small, about 5 mm long, initially brownish-red, later contrasted by the red-spotted white lip. The remaining petals are pressed together to form a helmet shape. Leaves are lanceolate.

Habitat and distribution
Widespread but local in England, on grassy hills and dry meadows. Prefers calcareous soil, poor in nutrients but warm and loamy or loess.

Orchid family *Orchidaceae*
Flowering time: May–June
Height of growth: 20–25 cm
Monocotyledonous; Perennial

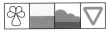

Identification marks
Inflorescence a lax spike 5–10 cm long containing 20–40 flowers. These are pale pink or pale purplish-red flushed with lilac. Lip is 1–1.5cm long, divided into 4 narrow lobes. Spur is short, thick, pointing downwards. Outer petals bend inwards to form a helmet shape, bearing conspicuous dark red veins. Lip has red spots.

Habitat and distribution
Semi-dry turf or meadows, also in dry thickets and dry open woodland on chalk and limestone.

Additional information
This is one of the rarest of British orchids, being restricted to one site each in Suffolk and Buckinghamshire.

Lady Orchid
Orchis purpurea

Orchis laxiflora
ssp. palustris

Orchid family *Orchidaceae*
Flowering time: May–June
Height of growth: 30–80 cm
Monocotyledonous; Perennial

Identification marks
Dense cylindrical spike 5–10 cm long containing more than 30 flowers. These are brownish-red, the lip is noticeably lighter, pink or almost white with numerous small red spots, 1.5–2 cm long and divided into 4 lobes, the front 2 being wider than the others. Spur is short, fat, pointing downwards. Outer petals bend inwards to form a helmet shape.

Habitat and distribution
Woodland with calcareous moist soil. Likes the warmth. Rare. Now restricted to a few sites in Kent. Abroad, it ranges across Europe to western Asia, but is always very local.

Orchid family *Orchidaceae*
Flowering time: May–July
Height of growth: 20–50 cm
Monocotyledonous; Perennial

Identification marks
Inflorescence a very lax spike with few flowers (5–15); spike grows to 5–10 cm long. Flowers dark purplish-red or light wine red. Outer petals spreading. Lip approx. 1.5 cm long and always wider than it is long, divided into 4 lobes. The 2 small inner lobes are somewhat longer than the lateral ones. Leaves narrow, 1.5–2 cm at the base, 10–20 cm long.

Habitat and distribution
Open moorland, moorland meadows. Likes warmth. Very rare.

Additional information
Similar: Jersey Orchid (*O. laxiflora* ssp. *laxiflora*), central lobes of lip shorter than the lateral ones. Western Europe including Channel Islands (but not mainland Britain), and the Mediterranean area. Rare.

Early Purple Orchid
Orchis mascula

Monkey Orchid
Orchis simia

Orchid family *Orchidaceae*
Flowering time: May–June
Height of growth: 20–50 cm
Monocotyledonous; Perennial

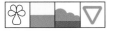

Identification marks
Spike 8–15 cm long, usually composed of many flowers. These are purple, pink and usually flushed with violet. Lip has 4 lobes and dark spots, with cylindrical spur which spreads horizontally. Outer petals projecting. Leaves are widest only in the middle, with or without darker spots.

Habitat and distribution
Semi-dry turf, poor unfertilized meadows and open thickets or deciduous woodland. Calcicolous. Scattered and locally common. Abroad, this species has a wide range through Europe and extends into north Africa and northern and western Asia.

Orchid family *Orchidaceae*
Flowering time: May
Height of growth: 20–40 cm
Monocotyledonous; Perennial

Identification marks
Inflorescence short, cylindrical or globular; starts flowering at the top and progresses downwards (other species flower in the reverse order). Flowers purplish-red or almost white. Lip 1–1.5 cm long, divided into 2 lateral sections and a middle section which is wide and curved with 2 long lobes. Spur is short, thick, pointing downwards. Lip is covered with fine red spots.

Habitat and distribution
Grassy slopes, and scrub in chalk hills of Kent and Oxford; rare. Occurs abroad in central and southern Europe, east to the Caucasus and south to north Africa.

Heath Spotted Orchid
Dactylorhiza maculata

Elder-flowered Orchid
Dactylorhiza sambucina

Orchid family *Orchidaceae*
Flowering time: June–July
Height of growth: 20–60 cm
Monocotyledonous; Perennial

Identification marks
Inflorescence a dense cylindrical to pyramid-shaped spike 4–8 cm long. Flowers light pink to almost white, with light purple patterns. Lip is trilobed, flat. Spur cylindrical, pointing downwards. Outer petals spreading, 5–7 mm long and thus smaller than the lip which grows to approx. 1 cm long.
Habitat and distribution
Damp woodland, heaths, open moorland, in peaty or acid soils. Scattered; where the plant occurs there are frequently colonies full of individual variations.
Additional information
The examples which grow on calcareous soil are usually classified as a separate species of Spotted Orchid (*D. fuchsii*).

Orchid family *Orchidaceae*
Flowering time: April–June
Height of growth: 15–25 cm
Monocotyledonous; Perennial

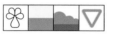

Identification marks
Short spike. In *D. sambucina* ssp. *sambucina* the spike is dense flowered, in *D. sambucina* ssp. *insularis* the flowers are loosely arranged. Outer petals spreading. Spur longer than ovary. Bracts leafy (not membraneous). Leaves elongate-ovoid, unspotted.
Habitat and distribution
Only cultivated in Britain. Widespread in Europe: meadows, scrub and light open woodland.
Additional information
The spike can vary from red through all the different shades to yellow. Yellow plants are usually more common.

Early Marsh Orchid
Dactylorhiza incarnata

Irish Marsh Orchid
Dactylorhiza majalis

Orchid family *Orchidaceae*
Flowering time: May–July
Height of growth: 20–60 cm
Monocotyledonous; Perennial

Identification marks
Inflorescence a dense cylindrical spike 4–15 cm long, the flowers flesh pink or light red. Lip is divided into 3 rather indistinct lobes. Spur cylindrical, pointing downwards. Outer petals spreading. 4–6 unspotted stem leaves. Lower and middle bracts of the inflorescence conspicuously longer than their respective flowers, upper ones usually as long as their flowers.

Habitat and distribution
Marshes, wet meadows, open wet moorland. Widely distributed.

Additional information
The Meadow Orchid comes in many different forms and flower colours. Occasionally cream or yellow specimens have been found.

Orchid family *Orchidaceae*
Flowering time: May–July
Height of growth: 10–50 cm
Monocotyledonous; Perennial

Identification marks
Inflorescence cylindrical-pyramidal, dense spike 5–10 cm long. Flowers lighter or darker shade of purplish red. Lip is trilobed. Spur cylindrical, pointing stiffly downwards. Outer petals spreading. Entire flower 1.5–2 cm long. 3–6 stem leaves always spotted. Bracts in inflorescence usually tinged with red, the lower ones longer than their flowers.

Habitat and distribution
Marshes, fens, wet meadows, wet moorland, in Ireland and western Scotland, not in England or Wales.

Additional information
Widespread in Europe, extending to central Russia, Siberia and the Caucasus. There it varies in stature, leaf shape and flower colour.

Fragrant Orchid
Gymnadenia conopsea

Orchid family *Orchidaceae*
Flowering time: May–Aug.
Height of growth: 15–60 cm
Monocotyledonous; Perennial

Identification marks
Inflorescence a cylindrical-pyramidal, dense spike which grows to 5–10 cm long. Flowers pink or pale purple with a tinge of violet, strongly scented. Lip wide, trilobed. Marginal petals horizontal and projecting. Spur thin and thread-like, at least 11/2 times as long as the ovary.

Habitat and distribution
Grassland, especially chalk and limestone, fens, marshes, woodland. Locally abundant.

Additional information
B. conopsea var. densiflora has bright rose-red or magenta flowers which smell of cloves. It is found only in Anglesey and the Isle of Wight.

Vanilla Orchid
Nigritella nigra

Orchid family *Orchidaceae*
Flowering time: May–Sept.
Height of growth: 5–25 cm
Monocotyledonous; Perennial

Identification marks
Inflorescence a compact, rounded spike. Flowers blackish-purple, approx. 0.5 cm long. Lip petal-like but conspicuously larger than the true petals. Innermost petals at most only half as wide as outer ones. Flowers smell very distinctly of vanilla. Leaves narrow, like grass; foliage rises up stem.

Habitat and distribution
Not British. Meadows in the Alps and Alpine foothills; also in the Pyrenees and Norway. Scattered.

Man Orchid
Aceras anthropophorum

Lizard Orchid
Himantoglossum hircinum

Orchid family *Orchidaceae*
Flowering time: May–June
Height of growth: 20–30 cm
Monocotyledonous; Perennial

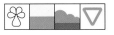

Identification marks
Inflorescence is a narrow spike usually 5–15 cm long. Flowers do not have spur. Outer petals come together to form helmet shape. They are green, have red or violet veins and margin. Lip approx 1 cm long, divided into narrow lobes.

Habitat and distribution
Grassland, scrub and woodland. Mainly on chalk and limestone. Thinly scattered though locally abundant in southern England.

Additional information
It is possible to see the deeply divided lip as the limbs of a hanging man (*anthropophorum* = carrying a man).

Orchid family *Orchidaceae*
Flowering time: April–June
Height of growth: 20–80 cm
Monocotyledonous; Perennial

Identification marks
Inflorescence a lax spike which grows 15–25 cm long. Outer petals light green, forming a helmet shape, often adhering along the margins. Conspicuous red veins. Lip trilobed: lateral lobes 5–7 mm long, usually brownish. Median lobe 5–7 cm long, 2–lobed at the apex.

Habitat and distribution
Grassland, open thickets, wood margins, mainly on chalk or limestone; rare.

Additional information
The species cannot be mistaken. In the Balkans there is the larger *H. calcaratum* with a lip narrower and more deeply divided.

Corydalis cava

Solid-tubered Fumitory
Corydalis solida

Fumitory family *Fumariaceae*
Flowering time: March–April
Height of growth: 10–20 cm
Dicotyledonous; Perennial

Identification marks
Stem erect, unbranched, with alternative deciduous leaves. These are bi- to tripinnate, bluish-green, glabrous. Ultimate segments are ovoid, often repeatedly dissected. 10–20 flowers form a dense raceme, bracts have entire margins. Flowers may be white to purple.

Habitat and distribution
A rare escape from gardens which may become established in a few places.

Additional information
Similar: Solid-tubered Fumitory (*C. solida*) (see right).

Fumitory family *Fumariaceae*
Flowering time: April–May
Height of growth: 10–20 cm
Dicotyledonous; Perennial

Identification marks
Stem erect, unbranched, with 2–3 alternate leaves. These are bluish-green, glabrous, double tripinnate. In the raceme there are 10–20 flowers, each in the axil of a bract. At least the lower bracts are divided into conspicuous finger-shaped lobes. Plants with reddish flowers found in the same locality as those with white flowers.

Habitat and distribution
Woodland, thickets. Local, often in colonies which have many different variations. A native of Europe, north and west Asia. Not native to Britain, but grown in gardens and naturalized locally.

Additional information
Similar: *C. intermedia*, only 1–5 flowers in a drooping inflorescence. Margin of bracts is entire.

Common Fumitory
Fumaria officinalis

Crown Vetch
Coronilla varia

Fumitory family *Fumariaceae*
Flowering time: May–Oct.
Height of growth: 10–40 cm
Dicotyledonous; Annual

Identification marks
10–50 flowers in a raceme 6–9 mm long. Flowers light purple, blackish-red at the apex. Leaves double pinnate. Apices of leaflets approx. 2 mm wide.

Habitat and distribution
Wasteland fields and gardens. Common.

Additional information
F. vaillantii, 6–12 flowers, 5–6 mm long, light pink. Calyx less than 1 mm long. Very local. *F. parviflora*, 1 10–20 flowers, 5–6 mm long. Flowers almost white, only red at the apex. Local.

Pea family *Fabaceae (Leguminosae)*
Flowering time: June–Sept.
Height of growth: 30–130 cm
Dicotyledonous; Perennial

Identification marks
10–20 flowers which grow to 1–1.5 cm long, sessile in a terminal umbel. They are purple, pink or white. The angular stems are prostrate or somewhat ascending. Leaves with long petioles, pinnate and have 11–25 leaflets. Each leaflet is on a very short petiole and is narrow-ovoid in shape.

Habitat and distribution
Grassland, pathways, open scrub. Scattered in Europe, but only locally naturalized in Britain.

Additional information
Crown Vetch survives on the poorest of soils. It improves soil, because the bacteria in its root nodules make nitrogen.

Red Clover
Trifolium pratense

Zigzag Clover
Trifolium medium

Pea family *Fabaceae (Leguminosae)*
Flowering time: June–Oct.
Height of growth: 15–45 cm
Dicotyledonous; Perennial

Identification marks
Usually 2 inflorescences at the end of each stem. They are rounded ovoid and are 2–3.5 cm long. Calyx half the length of the flower and has only 10 veins. Stem usually erect and hairy.

Habitat and distribution
Meadows and roadsides. Often grown in fields. Very common.

Additional information
Red Clover comes in many different varieties. These differ from one another in, among other things, the hairs on the stem, the length of the flowers and the flower colour. The cultivated races are also not uniform and there is much evidence of hybridization between wild and cultivated races.

Pea family *Fabaceae (Leguminosae)*
Flowering time: June–Aug.
Height of growth: 20–50 cm
Dicotyledonous; Perennial

Identification marks
Usually only 1 inflorescence at the end of each stem which is slightly rounded-ovoid and may become 2–4 cm long and 2–3 cm wide. Calyx glabrous, only the teeth of the calyx have conspicuous ciliate margin. Stem hairy, usually bent in a zigzag at each leaf node. Leaflets are 2–4 times longer than they are wide.

Habitat and distribution
Grassland, scrub, roadsides. Likes loamy soil. Common.

Additional information
Similar: Alsike clover (*T. hybridum*), more vigorous, to 60 cm, flower head globular, pink, carried on a short stem well above the uppermost leaf.

Trifolium rubens

Persian Clover
Trifolium resupinatum

Pea family *Fabaceae (Leguminosae)*
Flowering time: June–July
Height of growth: 30–60 cm
Dicotyledonous; Perennial

Identification marks
1–2 cylindrical inflorescences, 3–7 cm long
and 2 cm thick. Calyx usually glabrous on the
outside or covered with short hairs. Calyx
teeth have conspicuous ciliate margin. Stem
glabrous. Leaflets up to 6 cm long and 3–8
times longer than wide.
Habitat and distribution
Not British. Dry open woodland and scrub,
central Europe.
Additional information
Similar: Crimson Clover (*T. incarnatum*), 1
inflorescence up to 3–5 cm long, 1.5–2.5 cm
thick. Outside of calyx hairy. Leaflets max. 3
cm long and 1–1.5 cm wide. An annual native
to southern and western Europe: much
cultivated and locally naturalized.

Pea family *Fabaceae (Leguminosae)*
Flowering time: April–Sept.
Height of growth: 10–50 cm
Dicotyledonous; Annual

Identification marks
Inflorescence solitary on long peduncles in
the axils of the upper leaves. Each head 1–1.5
cm across. Flowers look odd initially as
individual flowers are twisted round 180° so
that the keel is facing upwards; reddish-violet
or pale red in colour, smelling distinctly of
honey.
Habitat and distribution
Not British. Path verges, roadsides, waste
ground. Sometimes cultivated.
Additional information
The wild form is smaller and is occasionally
considered a species in itself: *T. suaveolens*.

Common Sainfoin
Onobrychis viciifolia

Common Vetch
Vicia sativa

Pea family *Fabaceae (Leguminosae)*
Flowering time: May–June
Height of growth: 30–60 cm
Dicotyledonous; Perennial

Identification marks
Inflorescence a long lax raceme. Flowers 1–1.5 cm long, red or pink. Stem ascending or erect, with pinnate foliage. Leaves with 19–25 narrow leaflets.

Habitat and distribution
Well-drained grassland, roadsides, waste ground. Well scattered and appearing as wild in chalk and limestone grassland. Formerly used as a fodder crop and still sometimes planted for garden ornament.

Pea family *Fabaceae (Leguminosae)*
Flowering time: May–July
Height of growth: 30–80 cm
Dicotyledonous; Annual

Identification marks
1–2 flowers growing individually on short peduncles in leaf axils, each one 2–2.5 cm long. The standard is light purple or pink, wings carmine and darker than the standard, keel is whitish-pink. Stem more or less ascending. 8–12 leaflets, approx. 4–5 mm wide. The pinnate leaves usually terminate in a branched tendril.

Habitat and distribution
Roadsides, waste places, also planted as a fodder or green mature crop.

Additional information
Similar: *V. angustifolia*, smaller flowers and more prostrate, almost the same colour, only about 3 mm wide. Roadsides, dry grassland.

Narrow-leaved Everlasting Pea
Lathyrus sylvestris

Sea Pea
Lathyrus japonicus

Pea family *Fabaceae (Leguminosae)*
Flowering time: July–Aug.
Height of growth: 90–200 cm
Dicotyledonous; Perennial

Identification marks
3–10 flowers in a lax raceme, each 1.2–1.8 cm long, bright purplish-red. Stem prostrate, ascending or climbing, quadrangular, with 2 distinct wings. Leaves pinnate, composed of 2 leaflets and a branched tendril.

Habitat and distribution
Open woodland, scrub, hedgerows. Scattered and local.

Additional information
3 races are distinguished: typical subspecies (ssp. *sylvestris*) 3–6 flowers, leaflets 10–30 mm wide. Ssp. *angustifolius*, leaflets less than 5 mm wide. Scree slopes. Rare. Ssp. *platyphyllos*, leaflets 25–40 mm wide.

Pea family *Fabaceae (Leguminosae)*
Flowering time: June–July
Height of growth: 20–50 cm
Dicotyledonous; Perennial

Identification marks
3–10 flowers in a lax raceme. Keel is almost white, standard purple tending towards blue. Flowers approx. 1.5 cm long. Stem never winged, prostrate. 6–8 leaflets and branched tendril. Leaflets 2–4 cm long and half that width.

Habitat and distribution
Seaside beaches and dunes. Very local, from Suffolk to Kent, then scattered to Cornwall and south Wales. Also one locality in western Ireland.

Tuberous Pea
Lathyrus tuberosus

Black Pea
Lathyrus niger

Pea family *Fabaceae (Leguminosae)*
Flowering time: June–Sept.
Height of growth: 30–120 cm
Dicotyledonous; Perennial

Identification marks
1–5 flowers in a long-stalked raceme from the axils of the upper leaves. Flowers 1.5–1.8 cm long, brilliant carmine or purplish-red, occasionally tending towards violet. Leaves made up of a pair of leaflets and a branched tendril. Subterranean stem with tubers.
Habitat and distribution
Native to western Asia and Europe, in Britain only as a naturalized plant, occurring in hedgerows and very rarely in cornfields; rare.

Pea family *Fabaceae (Leguminosae)*
Flowering time: May–June
Height of growth: 30–80 cm
Dicotyledonous; Perennial

Identification marks
3–10 flowers in a lax cluster on a stalk from upper leaf axil. Flowers 1–1.5 cm long, purple to violet in colour, the actual petals often paler than the veins. Stem erect, branched without wings, angular. Leaves comprise 8–1 leaflets. When dried the entire plant goes black.
Habitat and distribution
Rocky woods on mountain slopes in Scotland rare, maybe extinct. Also naturalized in one site in Sussex.

Lathyrus liniifolius

Spring Pea
Lathyrus vernus

Pea family *Fabaceae (Leguminosae)*
Flowering time: April–June
Height of growth: 15–40 cm
Dicotyledonous; Perennial

Identification marks
3–6 flowers in a lax cluster which is situated in the leaf axil and may reach 7 cm in length. Flowers initially red, then dirty blue, 11–22 mm long. Stamens tubular in shape. Tube has a straight edge. Leaves comprise 4–6 leaflets whose undersides are usually distinctly bluish-green. Short point at leaf apex. Stem conspicuously winged even if this is normally only narrow.
Habitat and distribution
Not British. In Europe found in woodland, heaths, and mountain meadows.
Additional information
This plant is occasionally mistaken for Spring Pea (see p.333). It is, however, easy to recognize thanks to its winged stem.

Pea family *Fabaceae (Leguminosae)*
Flowering time: April–June
Height of growth: 20–60 cm
Dicotyledonous; Perennial

Identification marks
2–7 flowers in a lax cluster situated in a leaf axil; can grow to 6 cm long. Flowers initially red, then dirty blue, 1.5–2 cm long. Leaves made up of 4–6 leaflets approx. 1/2 as wide as long. Short point at leaf apex. Stem quadrangular and clearly without any wings.
Habitat and distribution
Not British. In Europe found in woodland, preferably deciduous. Prefers calcareous soil. Scattered, common locally.
Additional information
Spring Pea is occasionally confused with *Lathyrus liniifolius* (see p.333). It is however easy to recognize because the stem is not winged.

Alpine Sainfoin
Hedysarum hedysaroides

Tufted Milkwort
Polygala comosa

Pea family *Fabaceae (Leguminosae)*
Flowering time: July–Aug.
Height of growth: 15–50 cm
Dicotyledonous; Perennial

Identification marks
10–40 flowers in a somewhat 1-sided axillary raceme, up to 5–10 cm long. Each flower is 1.5–2 cm long, drooping. They are usually purple-red, more rarely creamy-coloured. Stem erect, unbranched, angular. Leaves pinnate, 9–19 leaflets. These grow to 1.5–3 cm in length and are lanceolate, sessile.

Habitat and distribution
Not British. Grassland, screes, stony slopes in the limestone Alps, more rarely in the central Alps and Pyrenees; uncommon. Sometimes cultivated in gardens.

Milkwort family *Polygalaceae*
Flowering time: May–June
Height of growth: 5–25 cm
Dicotyledonous; Perennial

Identification marks
10–30 flowers in a racemose inflorescence which is initially pyramidal in shape. The bracts beneath the flowers about as large as the newly-opened flowers (approx. 4 mm) and form a 'tuft' at the top of the relatively dense inflorescence shortly before opening. Flowers usually red, very rarely blue, occasionally lilac. Stem ascending or erect. Leaves alternate, spatulate to invert-ovoid.

Habitat and distribution
Not British. In Europe found in dry turf and poor meadows.

Additional information
Similar: Common Milkwort (*P. vulgaris*), bracts max. 2 mm long, and therefore not tufted. Native to chalk and limestone grassland in Britain, usually blue.

White Dittany, Burning Bush
Dictamnus albus

Indian Balsam
Impatiens glandulifera

Rue family *Rutaceae*
Flowering time: May–June
Height of growth: 40–120 cm
Dicotyledonous; Perennial

Identification marks
5-25 flowers in an erect terminal raceme, 2-3 cm across. Petals of unequal length. Basic colour pale pink. Veins deep red or reddish-violet. Stem erect. Leaves unpaired pinnate. 7-11 leaflets, short hairs, finely toothed and translucently spotted.

Habitat and distribution
Open scrub, forest margins. Prefers stony loose calcareous or loess ground. Native to southern Europe only. Cultivated in Britain.

Additional information
The plant contains ethereal oils. On hot days they evaporate so strongly that they can be ignited, hence the popular name 'Burning Bush'.

Balsam family *Balsaminaceae*
Flowering time: July–Oct.
Height of growth: 50–200 cm
Dicotyledonous; Annual

Identification marks
5-20 flowers in an erect axillary cluster; up to 2.5-4 cm long with short conspicuous spur pointing downwards. Stem is erect and usually unbranched. Leaves 10-25 cm long, sharply serrate.

Habitat and distribution
Wet woodland, river and lake banks. Originally a garden plant from the Himalayas, nowadays established widely.

Additional information
Similar: *I. balfourii*, 3-10 flowers per axillary raceme. Flowers are 2-coloured, white at the top, pink at the bottom. Plant rarely grows over 1 m high.

Wall Germander
Teucrium chamaedrys

Mint family *Lamiaceae (Labiatae)*
Flowering time: July–Sept.
Height of growth: 15–30 cm
Dicotyledonous; Perennial

Identification marks
2–6 flowers in the axis of the upper leaves all facing one way. Flower 1–1.5 cm long, fairly light reddish-violet in colour; no upper lip. Lower lip has 5 lobes. Stem ascending or erect, lower section woody. Leaves elongate cuneate, deeply toothed, rounded.

Habitat and distribution
Grassland, scrub, usually on calcareous soil. Central and southern Europe, not in the British Isles, but cultivated and locally naturalized.

Additional information
Similar: Water Germander (*T. scordium*), only 1–4 flowers (usually 3–4) in the axils of the upper leaves. Flowers 1-sided, less than 1 cm long, light purple, without upper lip. Stem not woody based; only in moist to wet places; very rare in Britain.

Bastard Balm
Melittis melissophyllum

Mint family *Lamiaceae (Labiatae)*
Flowering time: May–July
Height of growth: 20–50 cm
Dicotyledonous; Perennial

Identification marks
Stem quadrangular, erect, little branching. Broad leaves opposite and decussate, stalked, ovoid, conspicuously wrinkled, coarsely round-toothed margin. Entire plant has dense covering of soft hairs. Flowers few in number in axils of upper leaves, often all favouring one side; smells of honey.

Habitat and distribution
Light deciduous forests, woods and hedgerows. Prefers loose calcareous soils, warm but not too dry. Local to rare in Wales and southern parts of England.

Additional information
The flowers are usually white spotted with pink but may sometimes be more or less completely pink.

Common Hempnettle
Galeopsis tetrahit

Galeopsis pubescens

Mint family *Lamiaceae (Labiatae)*
Flowering time: July–Oct.
Height of growth: 10–80 cm
Dicotyledonous; Annual – Biennial

Identification marks
10–16 flowers in whorled sessile inflorescence terminating each stem. Flowers 1.5–2 cm long, their upper lip helmet-shaped. Lower lip has hollow tooth on each side. These are conical in shape and facing forwards. Central section of lower lip is more or less square.

Habitat and distribution
Weedy places on fields and wastelands, also in light open woodland areas. Common.

Additional information
Similar: *G. pubescens*, see right. Large-flowered Hempnettle (*G. speciosa*), flowers 2.5–3 cm long. Lower lip usually has bright violet spot on it (see p. 191).

Mint family *Lamiaceae (Labiatae)*
Flowering time: June–Oct.
Height of growth: 20–60 cm
Dicotyledonous; Annual – Biennial

Identification marks
10–16 flowers in whorled sessile inflorescence terminating each stem. Flowers 2–2.5 cm long, their upper lip helmet-shaped. Lower lip has a hollow tooth on each side, approx. 5 mm long pointing slightly forwards. Lower lip cropped at the front or slightly crenate, with dark violet and yellow spots. Stem slightly swollen at the nodes and clad with a mixture of long and short hairs.

Habitat and distribution
Not British. Light open woodland, fields. Scattered.

Additional information
Similar: Common Hempnettle (*G. tetrahit*), see left. Large-flowered Hempnettle (*G. speciosa*), flowers 2.5–3 cm long. Lower lip usually has bright violet spot (see p.191).

Black Horehound
Ballota nigra

Spotted Deadnettle
Lamium maculatum

Mint family *Lamiaceae (Labiatae)*
Flowering time: June–Aug.
Height of growth: 60–130 cm
Dicotyledonous; Perennial

Identification marks
4–10 flowers 1–1.5 cm long, on short peduncles in the axils of the upper and middle leaves; they are purplish-violet with white veins. Stem erect or ascending, hairy, branched, angular. Leaves in opposite pairs, heart-shaped to ovoid. Whole plant rank smelling.

Habitat and distribution
Waste places, roadsides, hedgerows and walls. Nitrogen indicator. Common.

Additional information
2 subspecies: that with leaves wider than 3 cm and calyx teeth which are over 4 mm is regarded as the typical subspecies and referred to as ssp. *nigra*. The leaves of the ssp. *foetida* are narrower than 3 cm.

Mint family *Lamiaceae (Labiatae)*
Flowering time: March–Oct.
Height of growth: 30–80 cm
Dicotyledonous; Perennial

Identification marks
Plant resembles nettle, but without stinging hairs. Many flowers in axils of middle and upper leaves, 2–3 cm long. Lower lip has large middle lobe, divided into 2 parts, and 2 small lateral lobes. Stem erect or ascending. Leaves opposite alternate, petiolate, ovate, 3–5 cm long, green or with a central white stripe.

Habitat and distribution
Woodland, scrub, roadsides, river banks. Likes damp soil rich in nutrients. Native to Europe, only naturalized locally in Britain, though much grown in gardens.

Additional information
Within the species many variations have been reported. The variations are to be found in virtually all the characteristics of the plant, especially size, leaf margin, hair covering.

Henbit
Lamium amplexicaule

Red Deadnettle
Lamium purpureum

Mint family *Lamiaceae (Labiatae)*
Flowering time: March–Oct.
Height of growth: 15–25 cm
Dicotyledonous; Annual

Identification marks
Plant resembles stinging nettle but without stinging hairs. 6–10 flowers in axillary whorls in upper leaf axils; up to 1–1.5 cm long; red, tending towards violet. The uppermost leaves are amplexicaul. The lower ones have long petioles and are opposite alternate. The stem is usually branched from the base.
Habitat and distribution
Fields, gardens, waste places, hedgerows, roadsides. Common.
Additional information
Some of the flowers of the Henbit do not open. The closed flowers can often be identified by their outstandingly bright red colour. Seed forms through self-pollination.

Mint family *Lamiaceae (Labiatae)*
Flowering time: March–Oct.
Height of growth: 10–40 cm
Dicotyledonous; Annual

Identification marks
Plant resembles stinging nettle but without stinging hairs. 3–5 flowers in whorls in the upper leaf axils. Flowers 1–2cm long; purple, usually with a strong shade of violet. Tip of the stem and the uppermost leaves usually also tinged with violet. Stem normally erect. All leaves are petiolate, opposite, alternate, ovate, wrinkled, crenate-toothed.
Habitat and distribution
Waste places, fields, roadsides and gardens. Very common.
Additional information
Cut-leaved Deadnettle (*L. hybridum*) has the appearance of a hybrid between Red Deadnettle and Henbit but is a true breeding species.

Betony
Betonica officinalis

Limestone Woundwort
Stachys alpina

Mint family *Lamiaceae (Labiatae)*
Flowering time: June–Aug.
Height of growth: 30–60 cm
Dicotyledonous; Perennial

Identification marks
Flowers are in a head-like terminal spike and in whorls in the axils of the uppermost leaves. Each flower is 1–1.5 cm long, pink or light purple, with a white spot towards the throat of the corolla. Basal leaves tufted with long petioles (stalk considerably longer than the leaf blade). Stem quadrangular, covered with coarse hairs like the leaves. Stem leaves petiolate, alternate, opposite, ovate, rounded-toothed.

Habitat and distribution
Grassland, often woodland, hedge-banks and heaths; common.

Additional information
Within the species one distinguishes the following groups: *B. hirsuta*, leaves have woolly layer on both sides. Rare. *B. stricta*, calyx has bearded teeth. Rare.

Mint family *Lamiaceae (Labiatae)*
Flowering time: June–Aug.
Height of growth: 50–100 cm
Dicotyledonous; Perennial

Identification marks
6–18 flowers in whorl-like clusters in the axils of terminal leaves. Stems erect. Flowers around 1.5 cm long or slightly longer, rather light flesh pink and a little dull. Margin of upper lip entire. Lower lip approx. twice as long as upper lip. Lower lip without markings. Leaves 5–20 cm long, pronouncedly serrate.

Habitat and distribution
Light woodland areas on calcareous soil rich in mull. Very rare. Absent in lowland areas.

Additional information
Limestone Woundwort is rare anywhere but in the Alps.

Hedge Woundwort
Stachys sylvatica

Marsh Woundwort
Stachys palustris

Mint family *Lamiaceae (Labiatae)*
Flowering time: June–Aug.
Height of growth: 60–120 cm
Dicotyledonous; Perennial

Identification marks
6–16 flowers in whorl-like inflorescences in upper leaf axils, forming terminal spikes with long bracts. The flowers are usually 1.2–1.5 cm long, purplish-red. The upper lip is shorter than the lower lip with distinct white markings.

Habitat and distribution
Woodland hedgebanks, field margins, waste ground; common.

Additional information
The pale forms may be taken for Limestone Woundwort. They are easily told apart: the lower lip of Hedge Woundwort always has white markings.

Mint family *Lamiaceae (Labiatae)*
Flowering time: June–Aug.
Height of growth: 10–60 cm
Dicotyledonous; Perennial

Identification marks
Inflorescences a whorled spike usually with 6 flowers. Flowers 1.5–1.8 cm long; purplish-red with a hint of violet. Upper lip is approx. 1/2 as long as the lower lip, which has light spots. Leaves 3–5 times longer than wide, oblong, lanceolate, crenate-toothed, 5–12cm long.

Habitat and distribution
Wet to damp grassland, fens and swamps; locally common.

Additional information
Abroad it has a wide range, Europe, temperate Asia east to Japan and North America.

Marjoram
Origanum vulgare

Breckland Thyme
Thymus serpyllum (sensu lato)

Mint family *Lamiaceae (Labiatae)*
Flowering time: July–Oct.
Height of growth: 30–60 cm
Dicotyledonous; Perennial

Identification marks
Numerous flowers in the upper leaf axils and at the end of the branches in head-like clusters in a compound panicle, approx. 5–6 mm long; pinkish-red to purplish-red. The bracts also have a reddish tinge. Leaves ovate, the margin often somewhat wavy or indistinctly toothed.

Habitat and distribution
Grassland, roadsides, scrub and forest margins. Prefers calcareous soil. Locally common on chalk and limestone formations.

Additional information
The species has many variations, these differing according to the number of flowers, their colour and the shape of the inflorescence.

Mint family *Lamiaceae (Labiatae)*
Flowering time: June–Oct.
Height of growth: 10–25 cm
Dicotyledonous; Perennial

Identification marks
Plant forms quite lax cushions. Flowers i head-like terminal inflorescence or sessile i whorls in upper leaf axils. Stem rounded quadrangular or indistinctly so, with hairs a round or with just 2 ribs of hairs.

Habitat and distribution
Sandy areas, short grassland, roadside heaths and screes; common.

Additional information
The most common Thyme species in Britai is *T. articus* ssp. *praecox* (syn. *T. drucei*). True *serpyllum* is only in East Anglia.

Field Cow Wheat
Melampyrum arvense

Red Bartsia
Odontites verna
(syn. rubra) ssp. *serotina*

Figwort family *Scrophulariaceae*
Flowering time: June–July
Height of growth: 10–30 cm
Dicotyledonous; Annual

Identification marks
Bracts pink, the flowers in their axils forming a dense terminal spike; purple, with a whitish tube and yellow throat, 2–2.5 cm long. Stem erect and usually branched; hairy.

Habitat and distribution
Weedy areas in corn fields in southern England, usually where the soil is limy. Very rare.

Additional information
20 years ago Cow Wheat was invariably present, though never common. As a result of the use of weedkillers it has virtually disappeared. Semi-parasite.

Figwort family *Scrophulariaceae*
Flowering time: Aug–Oct.
Height of growth: 10–40 cm
Dicotyledonous; Annual

Identification marks
Numerous flowers in upper leaf axils forming a rather one-sided spike; purplish pink; 1 cm long; hairy. Stem usually erect and begins branching at its base. The leaves are opposite, lanceolate and are widest in the basal half.

Habitat and distribution
Grassland, fields, waste places, especially where the soil is limy. Common.

Additional information
O. v. ssp. *verna*: usually approx. 20 cm high. Branches of the erect stem go off at conspicuously sharp angles; Flowering time June–July. All species of *Odontites* are semi-parasites.

Foxglove
Digitalis purpurea

Whorled Lousewort
Pedicularis verticillata

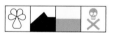

Figwort family *Scrophulariaceae*
Flowering time: June–Aug.
Height of growth: 60–180 cm
Dicotyledonous; Biennial

Identification marks
Inflorescence a long, wand-like, one-sided raceme. Flowers bell-shaped, somewhat lopsided, pendant, each one 3–5 cm long and almost 2 cm in diameter at the mouth. The throat is spotted. Basal leaves form a rosette. Leaves ovate, toothed and have grey, felt-like covering on the underside.

Habitat and distribution
Light woodland, heaths, roadsides and rocky slopes, mainly on acid soils. Locally common.

Additional information
Important medicinal plant. Contains numerous glycosides, the main one digitalin. Much grown in gardens and sometimes escaping to waste ground and pathsides.

Figwort family *Scrophulariaceae*
Flowering time: July
Height of growth: 5–20 cm
Dicotyledonous; Perennial

Identification marks
Numerous purplish-red flowers in a dense, clustered or head-shaped inflorescence approx. 1.5 cm long. Upper lip of each flower cropped at the apex; no teeth. Lower lip glabrous along the margin and projecting; it is half as long as the upper lip. Stem with 4 rows of hairs. Stem leaves in whorls of 3–4.

Habitat and distribution
Not British. Usually in limestone country between 1,300–2,000 m in damp grassland on screes. Widely spread in Europe, north to sub-Arctic Russia.

Additional information
Can be picked out immediately among the red-flowering species in this genus because of the whorled leaves. Semi-parasite.

Lousewort
Pedicularis sylvatica

Toothwort
Lathraea squamaria

Figwort family *Scrophulariaceae*
Flowering time: May–June
Height of growth: 5–15 cm
Dicotyledonous; Annual

Identification marks
5–15 flowers in a usually elongate terminal inflorescence. Flowers grow to 2–2.5 cm long, light pinkish-red. Upper lip rounded at the apex, with a tooth approx. 1 mm long on both sides. Lower lip shorter than upper lip. Stem prostrate, usually ascending at the tips. Leaves approx. 3 cm long, pinnate, glabrous, alternate.

Habitat and distribution
Needs damp, acid soil, usually on moorland and wet heaths. Widespread but local; commonest in Scotland.

Additional information
Stunted plants of the Marsh Lousewort (*P. palustris*) can occasionally resemble the Lousewort. It can be recognized by its branched, erect stems.

Figwort family *Scrophulariaceae*
Flowering time: March–May
Height of growth: 10–25 cm
Dicotyledonous; Perennial

Identification marks
The entire plant is without chlorophyll. It is either white, pink or flesh-coloured. Numerous flowers, 1–1.5 cm long, in a dense 1-sided raceme, nodding at the tip when young. Fleshy stem base carries pale-coloured scale-like leaves.

Habitat and distribution
In woodland and scrub; uncommon.

Additional information
A complete parasite, drawing nourishment from the roots of deciduous trees. Frequently found on Alder, Hazel or Poplar. In the right habitat Toothwort appears to live for many years.

Mezereon
Daphne mezereum

Striped Daphne
Daphne striata

Daphne family *Thymelaeaceae*
Flowering time: Feb.–April
Height of growth: 50–150 cm
Dicotyledonous

Daphne family *Thymelaeaceae*
Flowering time: May–July
Height of growth: 5–35 cm
Dicotyledonous

Identification marks
At flowering time plant has no leaves or leaf buds are just beginning to show green at tips. Flowers are pinkish-red to reddish-violet or rarely white; strong scent. Stem erect and fairly well branched, often with a strikingly wrinkled bark. When rubbed the bark has an unpleasant smell. The fruit ripens in midsummer and is a bright red drupe.

Habitat and distribution
Found in woodland, usually on chalk and limestone formations; rare.

Additional information
The plant contains the poisonous substance mezerine. On mucous membranes it can lead to unpleasant irritations.

Identification marks
An evergreen species, bearing 8–12 flowers in terminal clusters on branch tips. They are glabrous on the outside, pinkish-red and smell like lilac. The leathery, oblanceolate, glabrous leaves tend to be clustered at ends of branches, 1–2.5 cm long.

Habitat and distribution
Found wild only in the Alps (limestone Alps) on stony slopes and heaths where the soil is well drained. Sometimes cultivated in other parts of Europe, including Britain.

Additional information
Similar: *D. ptraea*: usually only 2–7 flowers in the terminal flower clusters. Flowers usually deeper shade of red. Only southern limestone Alps.

Rosemary Daphne
Daphne cneorum

Cranberry
Vaccinium oxycoccos

Daphne family *Thymelaeaceae*
Flowering time: May–June
Height of growth: 10–30 cm
Dicotyledonous

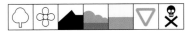

Identification marks
An evergreen species forming wide mats. 6–10 flowers in terminal clusters, with external thick layer of clinging hairs; deep pinkish-red; smell similar to carnations. Leathery leaves narrowly ablanceate and distributed evenly along the stem. Young stems covered with fine downy hairs.

Habitat and distribution
Found in grassland, among rocks and in open woodland, usually in limy soil in central Europe. Not native in Britain but widely cultivated.

Additional information
Cannot be confused outside the Alps, where it is easily distinguished from Striped Daphne by hairs on the branches and flower size and colour.

Heath family *Ericaceae*
Flowering time: June–Aug.
Height of growth: 10–80 cm
Dicotyledonous

Identification marks
Flowers solitary or in small groups up to 4 close to branch ends. Flowers small (approx. 5 mm across), their 4 or 5 petals reflexed, reddish-white to pink or deep pink. Peduncles conspicuously long and thin. Prostrate stems usually creep quite a distance across moss. Leaves normally evergreen, rolled up at the edges.

Habitat and distribution
Restricted to acid peat bogs, but there sometimes in abundance.

Additional information
Fruits contain large amount of vitamin C. They only taste pleasant after they have been subjected to frost.

Cross-leaved Heath
Erica tetralix

Erica carnea
(syn. *E. herbacea*)

Heath family *Ericaceae*
Flowering time: June–Sept.
Height of growth: 15–50 cm
Dicotyledonous

Identification marks
5–15 flowers in a terminal head which nods. Urn-shaped flowers approx. 6–8 mm long and flesh pink, with 4 small lobes at the apex. Branches erect, covered with white hairs. 3–4 evergreen leaves in a whorl; covered with stiff hairs.

Habitat and distribution
Bogs, wet heath and moorland; also in open woodland where the soil is moist and acid.

Additional information
Abroad it is widely spread in western and northern Europe, extending northwards to central Finland.

Heath family *Ericaceae*
Flowering time: Jan.–Aug.
Height of growth: 15–40 cm
Dicotyledonous

Identification marks
Usually more than 30 flowers in a conspicuous 1-sided terminal inflorescence; flesh pink; approx. 5–7 mm long, 4 small lobes at the apex. Needle-shaped, glabrous leaves usually in whorls of 4 along the stem.

Habitat and distribution
Not British. Native to the Alps, southern and central Europe and adjacent regions, on heaths and in light sub-Alpine woodland, usually between 1,500 and 2,200 m.

Additional information
Many cultivated varieties widely grown as ornamental plants.

Small-leaved Elm
Ulmus minor (syn. U. carpinifolia)

Wych Elm
Ulmus glabra

Elm family *Ulmaceae*
Flowering time: March–April
Height of growth: 5–35 m
Dicotyledonous

Elm family *Ulmaceae*
Flowering time: March–April
Height of growth: 20–40 m
Dicotyledonous

Identification marks
Flowers borne in dense clusters, appearing before the leaves, greenish-red with virtually no pedicel. The winged nutlet (samara) is glabrous. Wing obovate, with nutlet set off-centre. Seeds ripen and fall as the leaves expand. Leaves 4–10 cm long. Leaf blades obovate to oblanceolate, toothed and set very obliquely on the petiole.

Habitat and distribution
Widespread, but local, in Europe in lowland forest, but also much planted.

Additional information
Less common since the onset of Dutch elm disease. Similar: English Elm (*U. procera*), commonest Elm in Britain pre Dutch Elm disease. Still found in some areas. Leaves similar to *U. minor* but rough.

Identification marks
Flowers borne in dense clusters appearing before the leaves; greenish-red, virtually no pedicel. The winged fruit is glabrous; wing broadly elliptical, with nutlet set in the centre. Seeds ripen and fall as the leaves expand. Leaves 8–16 cm long. Leaf blades ovate, toothed and very harsh textured; borne obliquely on the stem. The 2 halves of each leaf are therefore very different from one another.

Habitat and distribution
Throughout Europe but local, often in hill or mountain woods. The only truly wild elm in Britain, but also much planted in the past. Susceptible, but less prone to Dutch Elm disease than *U. minor.*

Additional information
Similar: Small-leaved Elm and English Elm, see left.

Crab Apple
Malus sylvestris

Wild Cotoneaster
Cotoneaster integerrimus

Rose family *Rosaceae*
Flowering time: April–May
Height of growth: 2–10 m
Dicotyledonous

Identification marks
Twigs sometimes thorny. Leaves petiolate, broadly to almost round, often with off-centre apex. Leaf smooth, rarely more than 4 cm long, underside glabrous. Corymbs with few flowers. Anthers yellow, petals white or pink, 1–3 cm long. Fruit a small apple 2–3 cm across, dry and sour, somewhat woody.

Habitat and distribution
Deciduous woodland and light thickets, hedgerows. Likes calcareous, well-moistened soils rich in nutrients. Common in England and Wales, becoming rarer in Scotland.

Additional information
Very similar to the Apple and partially a hybrid of this is the cultivated Apple (*M. domestica*) with its many different varieties.

Rose family *Rosaceae*
Flowering time: May
Height of growth: 1–2 m
Dicotyledonous

Identification marks
2–4 somewhat bell-shaped flowers carried in nodding clusters of 2–4 in leaf axils. Flowers whitish, pink or purplish tinted, each one 5–8 mm long. Leaves alternate, margin entire, rounded to ovoid. 2–5 cm long, undersides covered with thick white felt, uppersides glabrous. Fruit pea-sized becoming blood red when ripe.

Habitat and distribution
In Wales only, very rare. Scattered through Europe on limestone hills among rocks.

Additional information
Similar: *C. tomentosus*. Usually only 1–2 flowers in leaf axils. Leaf undersides, sepals and fruit covered with felt-like hairs. Stony thickets on limestone. Very rare. Not British.

Dog Rose
Rosa canina

Rosa gallica
Rose family

Rose family *Rosaceae*
Flowering time: June
Height of growth: 1.3–3 m
Dicotyledonous

Identification marks
1–3 slightly scented flowers, 4–5 cm across; light pink or pale red. Calyx lobes reflexed when flower has faded. Stem thorny. Thorns usually hooked. Leaves pinnate. 5–7 leaflets, ovoid or elliptical, sharply serrate, glabrous.
Habitat and distribution
Deciduous woodland, thickets. Common.
Additional information
Within the genus there are several species hard to distinguish from one another (e.g. *R. squarrosa*). *R. obtusifolia*, leaf undersides glandular-hairy; thickets; rare. *R. dumetorum*, leaves hairy on the underside along the veins; thickets; scattered.

Rose family *Rosaceae*
Flowering time: June
Height of growth: 30–150 cm
Dicotyledonous

Identification marks
Flowers usually solitary, 6–7 cm across, bright red to dark purplish-red, inside at the base normally whitish. Peduncle stems covered with glands. Leaf margin glandular, often 5 leaflets only on flowering branches. Leaves smell slightly of vinegar.
Habitat and distribution
Not British. Forest margins, roadsides; scattered in Europe. Grown in gardens in Britain.
Additional information
Many suckers grow from the plant's roots so that in one habitat usually whole colonies grow up which, above the surface, seem to be independent.

Crowberry
Empetrum nigrum

Bog Rosemary
Andromeda polifolia

Crowberry family *Empetraceae*
Flowering time: May–July
Height of growth: 30–50 cm
Dicotyledonous

Identification marks
Flowers barely 3 mm long, inconspicuous, solitary in the axils of the upper leaves. Leaves needle-shaped, alternate or almost in whorls, very short petioles, glossy, undersides white, with rolled-under edges.

Habitat and distribution
Moors, heaths, dunes, alpine rocks. Needs acid soil. Usually abundant in those localities where it occurs.

Additional information
Two different races within the species: one has male and female plants. The male flowers pink, the female purple. The other race is hermaphrodite.

Heath family *Ericaceae*
Flowering time: May–Oct.
Height of growth: 15–30 cm
Dicotyledonous

Identification marks
1–5 flowers in a terminal inflorescence. Flowers globular to ovoid, campanulate, pink, 5–8 mm long. At the front tip of the bell are 5 small lobes. Narrow leaves, 3–5 mm wide, with rolled-under margins; upper-sides dark green with conspicuous veins, undersides light bluish-green. Stem erect or ascending.

Habitat and distribution
Bogs and wet heaths; scattered, local and decreasing, mainly in central Ireland and northern England to mid-Wales.

Additional information
The plant is losing habitats as bogs are drained and turned into farmland; contains poison, andrometoxin, just in its leaves or, more probably, in all its organs.

Cowberry
Vaccinium vitis-idaea

Bilberry
Vaccinium myrtillus

Heath family *Ericaceae*
Flowering time: June–Aug.
Height of growth: 10–30 cm
Dicotyledonous

Heath family *Ericaceae*
Flowering time: May–June
Height of growth: 15–40 cm
Dicotyledonous

Identification marks
Leaves leathery, evergreen; rolled up at the edge. Several flowers in terminal racemes, pink or pure white, slightly drooping. Flowers campanulate, usually 5, rarely 4 fused petals. Fruit: a berry, first white, then shining red when ripe, in dense clusters, usually unilateral.

Habitat and distribution
In mixed and coniferous woodland, high moorland, heaths with stunted bushy growth. Common in the mountains and sometimes becoming dominant. Requires acid, meagre soil saturated at intervals and containing coarse humus.

Additional information
Similar: Bearberry (*Arctostaphylos uva-ursi*), especially when this is not in flower. The leaf edges are flat (p. 114).

Identification marks
Axillary flowers solitary; globular-campanulate, greenish and usually tinged with red. As a rule they have 5, occasionally 4 petal-lobes. Leaves deciduous, ovate, pointed, margin slightly rounded-toothed, green on both sides. Stem angular, green. Berries blue-black.

Habitat and distribution
Moors, heaths, acid woods, usually on hills or mountains. Usually abundant in the localities where it occurs. Common, except for eastern England from the Humber to the Thames.

Additional information
Similar: Bog Bilberry (*V. uliginosum*), leaves obovate to oval, without teeth, the undersides bluish-green. Stems brown. Moors and peaty forests. Scattered in west Scotland; rare in extreme north of England.

Hairy Alpenrose
Rhododendron hirsutum

Rusty Alpenrose
Rhododendron ferrugineum

Heath family *Ericaceae*
Flowering time: May–Aug.
Height of growth: 50–120 cm
Dicotyledonous

Identification marks
Flowers in a terminal umbel, funnel-shaped to campanulate, approx. 1.5 cm long, with 5 lobes, light red, hairy on the inside. Leaves elliptic evergreen, leathery, light green and glossy on the upperside, matt on the underside and spotted initially with yellow, then with brown glandular scales. The entire leaf margin covered with conspicuous layer of ciliate hairs.

Habitat and distribution
Only cultivated in Britain. In central and eastern Alps, 2,400 m. Scattered and locally abundant where it occurs.

Additional information
Similar: Rusty Alpenrose (*R. ferrugineum*), see right.

Heath family *Ericaceae*
Flowering time: May–Aug.
Height of growth: 50–200 cm
Dicotyledonous

Identification marks
Flowers in a terminal umbel, funnel-shaped to campanulate, approx. 1.5 cm long, with 5 lobes, dark red, hairy on the inside. The narrowly oblong to elliptic leaves evergreen, leathery, dark green on the top, the undersides covered with a dense layer of rusty yellow glandular scales. Leaf margin rolled under and never has ciliate hairs.

Habitat and distribution
Cultivated in Britain. Alps, Pyrenees and the Jura in peaty acid soils.

Additional information
Similar: Hairy Alpenrose (*R. hirsutum*): leaves have conspicuous ciliate hairs, limestone Alps.

Heather
Calluna vulgaris

Heath family *Ericaceae*
Flowering time: July–Sept.
Height of growth: 20–50 cm
Dicotyledonous

Identification marks
Flowers in terminal racemes, sometimes branched. Each flower 2–4 mm long and deeply cleft in 4 lobes; pink to purplish-red, shorter than the petal-like 4-lobed calyx of the same colour. Leaves in 4 rows on the stem, scale-like.

Habitat and distribution
Found on heaths, moors, bogs and in open woods where the soil is acid; common.

Additional information
Heather does not decay easily. Crude humus is formed from its dead remains.

Spiny Restharrow
Ononis spinosa

Pea family *Fabaceae (Leguminosae)*
Flowering time: June–Sept.
Height of growth: 30–60 cm
Dicotyledonous

Identification marks
Flowers axillary in a loose raceme on the upper part of the stem. Stem with 2 lines of hairs only, erect or ascending, often thorny. Leaves made up of 3 finely toothed leaflets.

Habitat and distribution
Dry pastures, roadsides and waste ground; scattered and local.

Additional information
2 groups within the species: Spiny Restharrow and Creeping Restharrow (often listed as *O. repens*). Creeping Restharrow has no thorns; is not rhizomatous, and its stem is evenly covered with hairs.

Columbine-leaved Meadow Rue
Thalictrum aquilegifolium

Round-leaved Penny-cress
Thlaspi rotundifolium

Buttercup family *Ranunculaceae*
Flowering time: May–June
Height of growth: 50–150 cm
Dicotyledonous; Perennial

Identification marks
Lilac to purple or white flowers consist of virtually only the bushy projecting stamens which are arranged in a rather dense panicle. Leaves alternate on the stem, bi- or tripinnate with rounded leaflets which are crenate at the apex.

Habitat and distribution
Damp woodland, river banks, mountainsides, eastern and central Europe.

Additional information
Only cultivated in Britain. In the past the plant was used to dye cloth; leaves contain a yellow dye.

Mustard family *Brassicaceae (Cruciferae)*
Flowering time: July–Sept.
Height of growth: 5–15 cm
Dicotyledonous; Perennial

Identification marks
Stem creeping, producing erect or ascending flowering shoots which bear many leaves. Leaves glaucous, ovoid, margin entire or serrate. Flowers in compressed corymbs. Fruit oval, somewhat flattened, approx. twice as long as wide.

Habitat and distribution
Not British. Only in Alpine areas over 1,000–1,500 m.

Additional information
Prevalent form is distinguished by its blue-tinged corolla with darker veins. White-flowered form is rare.

Dame's Violet
Hesperis matronalis

Perennial Honesty
Lunaria rediviva

Mustard family *Brassicaceae (Cruciferae)*
Flowering time: May–July
Height of growth: 40–100 cm
Dicotyledonous; Biennial – Perennial

Identification marks
Numerous flowers in erect open racemes; violet or white; up to 2 cm across. The ripe pods measure up to 9 cm. Basal leaves ovoid, usually wither during later part of flowering period; up to 15 cm long. The stem leaves hairy.
Habitat and distribution
Cultivated in Britain. Forest margins, roadsides and waste places, usually on somewhat moist soil. Native to Europe, western and central Asia locally.

Mustard family *Brassicaceae (Cruciferae)*
Flowering time: May–July
Height of growth: 30–150 cm
Dicotyledonous; Perennial

Identification marks
10–30 flowers; approx. 1.5 cm across; usually pale violet, sometimes whitish-violet. Ripe fruits are particularly conspicuous: 3–5 cm long, in rare cases to even 9 cm and approx. 1/3 as wide as they are long. Leaves petiolate with deep heart-shaped lobes, toothed.
Habitat and distribution
Cultivated in Britain. Woodland on humus, stony and damp ground. Throughout Europe.
Additional information
Grown in borders of perennial flowers but not frequently seen. Common Honesty (*L. annua*) is more frequently seen, both in gardens and naturalized by roadsides and on waste ground.

Cuckoo Flower
Cardamine pratensis

Coralroot
Cardamine bulbifera

Mustard family *Brassicaceae (Cruciferae)*
Flowering time: April–June
Height of growth: 15–60 cm
Dicotyledonous; Perennial

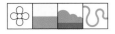

Identification marks
Basal leaves form rosette, pinnate. Leaflets rounded, terminal leaflet usually larger. Stem leaves pinnate with narrow tips. Hollow stem. Inflorescence a raceme. Flowers large (approx. 1–1.5 cm across), petals longer than sepals. Fruit much longer than wide. Flower colour depends on habitat: white (shady), pink, mauve or deep purple (dry).

Habitat and distribution
Mainly in damp meadows and pastures on loamy soil. Indicator of rich ground and ground water. Common.

Additional information
Similar: Narrow-leaved Bitter-cress (*C. impatiens*), flowers only 0.5 cm across, petals as long as sepals, whitish; fruit bursts if touched; woods.

Mustard family *Brassicaceae (Cruciferae)*
Flowering time: April–May
Height of growth: 30–70 cm
Dicotyledonous; Perennial

Identification marks
Flowers usually violet, sometimes pink or white. No basal rosette of leaves. Leaves at least partially pinnate. Upper leaves narrow and entire. Dark brownish-purple bulbils in leaf axils.

Habitat and distribution
Deciduous and mixed woodland in hilly areas on soil rich in nutrients and usually calcareous. Local in southern England Midlands to southern Scotland.

Additional information
Coralroot reproduces mainly by means of its bulbils as seeds are rarely formed. Ants carry off the bulbils, so often only small numbers in each locality. Synonym: *Dentaria bulbifera*.

Cross Gentian
Gentiana cruciata

Fringed Gentian
Gentianella ciliata

Gentian family *Gentianaceae*
Flowering time: July–Oct.
Height of growth: 10–40 cm
Dicotyledonous; Perennial

Identification marks
Flowers have short peduncles or none at all, arising in the upper leaf axils and sometimes lower down, either solitary or up to 3 in a group. Flowers 2–2.5 cm long, campanulate and erect, with 4, more rarely 5 blunt lobes, cleft to approx. 1/3 of their length. Leaves opposite and lanceolate with leathery appearance.

Habitat and distribution
Not British. Mainly in southern, central and eastern Europe, north to Holland. Thickets, forest margins and grassland. Very rare.

Additional information
Species very much reduced by fertilizers on grassland where it used to be found.

Gentian family *Gentianaceae*
Flowering time: Aug.–Oct.
Height of growth: 10–25 cm
Dicotyledonous; Biennial – Perennial

Identification marks
Stem usually with one terminal flower, more rarely 2–10. Flowers 2–5 cm long, cleft to about 1/2 into 4 lobes. Lobes have long fringe particularly at the base. Flower deep blue, more rarely pale blue. Margin of petal lobes often slightly rolled. Leaves linear-elongate with a single vein.

Habitat and distribution
Not British. Grows in light open woodland and grassland on limestone formations.

Additional information
Easiest Gentian to recognize in Europe; not common anywhere, nevertheless widely distributed. No subspecies known.

Corn Mint
Mentha arvensis

Spearmint
Mentha spicata

Mint family *Lamiaceae (Labiatae)*
Flowering time: June–Oct.
Height of growth: 15–50 cm
Dicotyledonous; Perennial

Identification marks
All flowers in axillary whorls in upper 6–10 leaf pairs. Stem tip leafy. Leaves elongate, more rarely rounded, always toothed and hairy, opposite, alternate. Stem quadrangular.
Habitat and distribution
River banks, ditches, wet areas in fields, meadows, woodland. Common in the south, less so in the north and Scotland.
Additional information
It is difficult to differentiate clearly between the subspecies. Also hybrids of the Corn Mint and other species within the genus are by no means rare.

Mint family *Lamiaceae (Labiatae)*
Flowering time: July–Aug.
Height of growth: 30–90 cm
Dicotyledonous; Perennial

Identification marks
Inflorescence a slender, whorled spike, often branched. Floral bracts very narrow, almost like bristles. Flowers blue-lilac, 3 mm wide. Leaves 6–10 cm long, 2–3 cm wide, almost sessile, never wrinkled.
Habitat and distribution
River banks, ditches, damp meadows, roadsides, waste ground. Native to central Europe only, but widely naturalized elsewhere, including the British Isles. Much cultivated as a herb for flavouring.
Additional information
Similar: Water Mint (*M. aquatica*), terminal inflorescence dense, rounded; beneath it 1–2 axillary flower whorls. River, stream and pond banks, marshes and fens. Common.

Brooklime
Veronica beccabunga

Blue Water Speedwell
Veronica anagallis-aquatica

Figwort family *Scrophulariaceae*
Flowering time: May–Aug.
Height of growth: 20–60 cm
Dicotyledonous; Perennial

Identification marks
Usually 10 flowers (occasionally more) in a lax axillary raceme. Flowers 6–8 mm across, divided into 4 blue petal lobes. Leaves opposite, glabrous, glossy, petiolate and notched, oval to rounded.
Habitat and distribution
Ponds, streams, rivers, marshes and wet grassland; common.
Additional information
Plants within this species may vary considerably. Submerged plants look very different. Water Speedwell (see right) always has leaves more than twice as long as wide.

Figwort family *Scrophulariaceae*
Flowering time: June–Oct.
Height of growth: 15–50 cm
Dicotyledonous; Perennial

Identification marks
Usually 20–50 flowers in a dense axillary raceme, 4–6 mm across and divided into 4 pale blue petal lobes. Leaves opposite, usually glabrous, sessile or with short petioles, semi-ampexicaul, lanceolate, up to 2 cm long.
Habitat and distribution
Rivers, streams, ponds and wet meadows; fairly common.
Additional information
In Europe this species has many varieties, but in Britain the regional populations are fairly uniform.

Ivy-leaved Speedwell
Veronica hederifolia

Grey Field Speedwell
Veronica polita

Figwort family *Scrophulariaceae*
Flowering time: March–May
Height of growth: 5–30 cm
Dicotyledonous; Annual

Identification marks
Flowers solitary in leaf axils, 2–5 mm across, blue or bluish-violet, more rarely white. Stem prostrate or ascending, branched. Leaves 3–7 lobed, ciliate and slightly hairy.

Habitat and distribution
Hedgerows, waste ground, fields and gardens, occasionally in forest clearings. Very common.

Additional information
Similar: *V hederifolia*, ssp. *sublobata* has fruit stems longer than 4 times the length of the calyx. Ssp. *hederifolia* has fruit stems twice as long as the calyx.

Figwort family *Scrophulariaceae*
Flowering time: March–Sept.
Height of growth: 5–20 cm
Dicotyledonous; Annual

Identification marks
Flowers solitary in leaf axils, 4–8 mm across, dark blue. Stem prostrate (but not rooting), ascending or erect. Leaves 0.5–1.2 cm long, rounded, dark green, somewhat glossy. Fruit stem at maturity as long as the leaves or shorter.

Habitat and distribution
Gardens. Common.

Additional information
Similar: Field Speedwell (*V. agrestis*), leaves conspicuously longer than they are wide. Fruit stem at least 1½ times as long as the leaves. Flowers pale blue and white. Weedy areas, cultivated ground. Local in the south, common in the north.

Common Field Speedwell
Veronica persica

Fingered Speedwell
Veronica triphyllos

Figwort family *Scrophulariaceae*
Flowering time: March–Dec.
Height of growth: 10–40 cm
Dicotyledonous; Annual – Biennial

Identification marks
Flowers solitary in leaf axils, 8–12 mm in diameter, sky blue with whitish or yellowish spot in the throat. Stem prostrate or ascending. Leaves heart-shaped or ovate, coarsely toothed, minutely ciliate and hairy on the veins beneath.

Habitat and distribution
Waste ground, fields and in gardens. Common.

Additional information
Plant is a native of western Asia. It must have spread via Turkey to the Balkans around 1800 and from there to the south-eastern part of central Europe. First recorded in Britain 1825.

Figwort family *Scrophulariaceae*
Flowering time: March–May
Height of growth: 5–15 cm
Dicotyledonous; Annual

Identification marks
Flowers in a short, open raceme; 5–7 mm across, dark blue. Flower stems longer than the calyx. Stem erect, lower part branched. Middle and upper leaves 3–5 lobed, lower ones ovoid.

Habitat and distribution
Sandy fields in East Anglia, very rare.

Additional information
Similar: *V. praecox*, flowers barely over 5 mm across. Leaves undivided. Sandy fields in East Anglia. Rare. *V. acinifolia*, flowers only 3–5 mm across, light blue. Middle and upper leaves undivided. Damp muddy fields. Very rare.

Wall Speedwell
Veronica arvensis

Figwort family *Scrophulariaceae*
Flowering time: March–Sept.
Height of growth: 5–20 cm
Dicotyledonous; Annual

Identification marks
Flowers in erect racemes. They are light blue, with very short petioles and only grow to 4–5 mm in diameter. Leaves toothed, lower ones heart-shaped to ovate, upper ones narrower usually with margin entire.

Habitat and distribution
Waste ground, fields, gardens, heaths and grassland; common.

Additional information
V. verna, flowers rich blue, 3–4mm in diameter. Middle leaves pinnate, lower ones ovate and barely toothed. Plant 5–10 cm high. Sandy fields in East Anglia; rare. *V. dillenii*, flowers dark blue, 5–7 mm in diameter. Middle leaves pinnate, lower ones ovate. Sandy ground, but not in Britain.

Heath Speedwell
Veronica officinalis

Figwort family *Scrophulariaceae*
Flowering time: June–Aug.
Height of growth: 15–30 cm
Dicotyledonous; Perennial

Identification marks
15–25 flowers in fairly axillary racemes approx. 6 mm across, usually lilac, sometimes very pale or almost white. Stem creeping with ascending flowering stems. Leaves opposite, hairy, leathery, toothed, with short petioles.

Habitat and distribution
Open woodland, heaths and grassland usually on somewhat acid ground. Common.

Additional information
The species is very evenly distributed Nevertheless there are virtually no reports of any varieties other than the occasional occurrence of white or pink flowered specimens.

Germander Speedwell
Veronica chamaedrys

Veronica austriaca

Figwort family *Scrophulariaceae*
Flowering time: April–June
Height of growth: 15–30 cm
Dicotyledonous; Perennial;

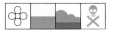

Identification marks
10–30 flowers in racemes in axils of upper leaf pairs. Flowers 10 mm across, bright blue with darker veins. Stem with normally 2 conspicuous rows of hairs. Leaves opposite, short petioles or sessile, up to 3.5 cm long, almost twice as long as wide, ovate, toothed and hairy.

Habitat and distribution
Grassland, open woodland, hedgerows and roadsides. Common.

Additional information
Similar: *V. urticifolia*, usually only 4 lax clusters. Flowers 6–8 mm across. Stem evenly covered with hairs or glabrous. Leaves up to 10 cm long. Woodland. Scattered. Not British.

Figwort family *Scrophulariaceae*
Flowering time: May–Aug.
Height of growth: 15–50 cm
Dicotyledonous; Perennial

Identification marks
Flowers in dense axillary racemes at the top of the stems. Flowers 1–1.5 cm across, dark blue. Stem ascending or erect, covered with curly hairs. Leaves ovate to oblong-lanceolate, sessile, or lower ones with short petioles.

Habitat and distribution
Scrub, woodland and grassland. Scattered throughout Europe. Much cultivated.

Additional information
The species currently comprises various groups. The one described above is most common, once classified as *V. teucrium*.

Hoary Plaintain
Plantago media

Field Madder
Sherardia arvensis

Plaintain family *Plantaginaceae*
Flowering time: May–June
Height of growth: 15–30 cm
Dicotyledonous; Perennial

Madder family *Rubiaceae*
Flowering time: May–Oct.
Height of growth: 5–30 cm
Dicotyledonous; Annual

Identification marks
Stem 2–5 times as long as flower spike. Spike up to 8 cm long, dense. Flowers inconspicuous. Stamens lilac, on long purple filaments. Leaves in a rosette, margins slightly toothed. Leaf blade at least 4 times as long as petiole.

Habitat and distribution
Dry grassland, meadows, roadsides, usually on limy soils; widespread but local, less common in the north.

Additional information
Similar: *P. intermedia*, stem twice as long as flower spike. Spike slender. Stamens short and only initially pale lilac. Leaf blade max. twice as long as petiole. Leaves normally flat on the ground. Damp fields and roadsides. Not British.

Identification marks
Flowers in a terminal, rounded, false umbel of few flowers, each 3–4 mm across. Prostrate or ascending stem conspicuously quadrangular. Leaves arranged in whorls of 5–6 on the upper parts of the stem, but only in 4s at the base. Leaves have single vein, rough along the margins.

Additional information
Fields, grassland, roadsides, waste ground; widely spread but local.

Additional information
Field Madder never very common, but until the advent of chemical herbicides it was never absent in any area where the soil conditions were suitable.

Teasel
Dipsacus fullonum

Devil's-bit Scabious
Succisa pratensis

Scabious family *Dipsacaceae*
Flowering time: July–Aug.
Height of growth: 90–200 cm
Dicotyledonous; Biennial

Identification marks
Flowers in a dense thimble-shaped head with protruding, bristle-like bracts. Flowers lilac, opening in rings from the middle upwards and downwards. Bracts of the capitulum bearing small spines and facing upwards. Margin of stem leaves entire or toothed. Stems erect, ribbed and prickly.
Habitat and distribution
Roadsides, hedgerows, waste ground and grassland. Widespread but local, commoner in the south.
Additional information
Similar: *D. laciniatus*, bracts of the capitulum project horizontally or downwards, without thorns. Stem leaves irregularly pinnate. Not native in Britain.

Scabious family *Dipsacaceae*
Flowering time: July–Aug.
Height of growth: 90–200 cm
Dicotyledonous; Biennial

Identification marks
Flowers in dense hemispherical heads, 1.5–2.5 cm across; mauve to violet-blue. Marginal flowers no larger than inner ones. Between flowers conspicuous black bristles (pull out flowers to see). Stem below capitulum covered with clinging hairs. Leaves opposite, ovoid-lanceolate, undivided.
Habitat and distribution
Damp meadows, open moorland, damp open woodland. Scattered.
Additional information
Similar: *S. inflexa*, flowers lilac. There are no black bristles between the flowers. Very rare in open moorland. Not British; predominantly in eastern Europe.

Dianthus superbus

Columbine
Aquilegia vulgaris

Pink family *Caryophyllaceae*
Flowering time: June–Sept.
Height of growth: 30–90 cm
Dicotyledonous; Perennial

Identification marks
Flowers solitary, on pedicels. Petals lilac to deep pink, often with strong tendency towards lilac. Petals arching, fringed, divided beyond the middle, often with dark spots. Flower generally over 2.5 cm across. Short epicalyx scales at base of calyx no more than 1/3 of calyx length. Stem leaves opposite, often somewhat bluish-green, 3–10 mm wide. Flowers usually have strong scent.

Habitat and distribution
Cultivated in Britain. Open woodland, scrub, mountain and wet meadows in much of Europe.

Additional information
Similar: Pink (*D. plumarius*), flower opening to 2.5 cm across, pink to white. Not British. Eastern central Europe.

Buttercup family *Ranunculaceae*
Flowering time: June–July
Height of growth: 30–60 cm
Dicotyledonous; Perennial

Identification marks
3–12 flowers, each 3.5–5 cm across in very open panicle. 5 petals with spurs. Spur erect, bent into a hook at the end. Leaves doubly divided into 3.

Habitat and distribution
Damp woodland, fens, wet meadows; native and naturalized; local and rare.

Additional information
Similar: Purple Columbine (*A. atrata*), flowers brownish-violet in colour. Stamens extend 1 cm beyond unfolded flower. Woodland, meadows. *A. einseleana*, flower 2–4 cm across. Often only 1 flower, max. 6. Spur barely curved: eastern limestone Alps. Rare. Alpine Columbine (*A. alpina*), flower diameter 5–8 cm. Spur barely curved at tip. 1–3 flowers. Alps. Rare. None native to Britain.

Meadow Crane's-bill
Geranium pratense

Wood Crane's-bill
Geranium sylvaticus

Geranium family *Geraniaceae*
Flowering time: June–Sept.
Height of growth: 30–60 cm
Dicotyledonous; Perennial

Identification marks
Inflorescence several axillary peduncles from top of erect stems each having 2 flowers. Flowers 2.5–4 cm across, blue, sometimes with violet undertone. Peduncles bend downwards after flowering. Leaves large, palmate, with 7 toothed lobes.

Habitat and distribution
Meadows, roadsides. Scattered, common locally, but not in northern Scotland.

Additional information
The Meadow Crane's-bill is distributed predominantly in eastern Europe. In meadows with rich soil, it can occur in such abundance that its blue flowers dominate the scene. It is rarely so abundant in Britain.

Geranium family *Geraniaceae*
Flowering time: June–Sept.
Height of growth: 30–60 cm
Dicotyledonous; Perennial

Identification marks
Inflorescence several axillary peduncles from tops of erect stems each having 2 flowers. Flowers 2–3 cm across, reddish or bluish-violet. Leaves palmate, 7–12 cm wide, usually divided into 7 segments, divisions never as deep as the stem.

Habitat and distribution
Open woodland, meadows, hedge-banks, mountain rocks. Well distributed in northern England and Scotland, rare or absent elsewhere. Where it occurs it can be in large numbers.

Additional information
Some Wood Crane's-bill vary in petal colour.

Sea Holly
Eryngium maritimum

Blue Pimpernel
Anagallis foemina

Umbellifer family *Apiaceae (Umbelliferae)*
Flowering time: June–Oct.
Height of growth: 10–40 cm
Dicotyledonous; Perennial

Identification marks
Inflorescence a compact hemispherical terminal umbel which elongates after flowers have withered. Flowers small, blue, tending towards violet. Involucral bracts spread out like stars, 2–4 cm long, wavy and with spiny teeth. Basal leaves rounded or reniform, also with teeth with long, spiny tip. Stem richly branched. Plants often form semi-spherical 'bush'. Whole plant is tinged with blue-grey.

Habitat and distribution
Dunes, shingle beaches, always near the sea, around whole of Britain.

Primrose family *Primulaceae*
Flowering time: June–Sept.
Height of growth: 10–20 cm
Dicotyledonous; Annual

Identification marks
Single axillary flowers approx. 5 mm across; bluish-violet inside, blue outside. Petals do not overlap, front margin conspicuously toothed. Stem prostrate. Leaves opposite, 0.5–2 cm long, narrow ovoid.

Habitat and distribution
Fields and waste ground; rare.

Additional information
Syn: *A. arvensis* spp. *foemina*. Similar: Scarlet Pimpernel, *A. arvensis*, flowers red, rarely mauve or blue. Petals touch each other or overlap, never toothed at the apex. Fields, gardens, waste places, roadsides. Common.

Sea Lavender
Limonium vulgare

Lesser Periwinkle
Vinca minor

Sea Lavender family *Plumbaginaceae*
Flowering time: July–Sept.
Height of growth: 20–50 cm
Dicotyledonous; Perennial

Identification marks
Numerous violet flowers in dense cymes, in stiff, panicle-like inflorescence. Each flower 6–8 mm long, with 5 petal lobes spread out in semi-campanulate fashion; usually conspicuously crenate. All leaves in a basal rosette; 5–20 cm long, 1.5–3 cm wide.

Habitat and distribution
Salt marshes, scattered around the coasts of England and Wales but not Scotland.

Additional information
Varieties growing on beaches of the Mediterranean, Atlantic, North Sea and Baltic differ somewhat from one another.

Dogbane family *Apocynaceae*
Flowering time: April–May
Height of growth: 10–20 cm
Dicotyledonous; Perennial

Identification marks
Flowers solitary, light blue-purple, and occasionally white, 2–3 cm across. Petal lobes spread out flat. Stem trailing, woody at the base. Flowering stems ascending. Leaves opposite, lanceolate, leathery, evergreen, glabrous, up to 4 cm long.

Habitat and distribution
Woodland, hedgebanks and roadsides. Scattered. Not considered native to Britain.

Additional information
Frequently cultivated as is Greater Periwinkle (*V. major*), flower 4–5 cm across. Plant grows to 50 cm. Leaves have ciliate margin, up to 10 cm long. Less hardy than *V. minor*.

Marsh Gentian
Gentiana pneumonanthe

Milkweed Gentian
Gentiana asclepiadea

Gentian family *Gentianaceae*
Flowering time: July–Sept.
Height of growth: 10–40 cm
Dicotyledonous; Perennial

Identification marks
Usually 1–3 flowers at stem end, occasionally up to 7; 3.5–5 cm long; 5 green lines on outside. Stem erect, unbranched. Leaves opposite, narrow and blunt.
Habitat and distribution
Wet heathland in England and Wales. Local and decreasing with loss of suitable habitats.
Additional information
Similar: Milkweed Gentian (*G. asclepiadea*), numerous flowers either single or in 2s and 3s usually in upper leaf axils.

Gentian family *Gentianaceae*
Flowering time: June–Sept.
Height of growth: 40–70 cm
Dicotyledonous; Perennial

Identification marks
Flowers terminal, in the axils of middle and upper leaves, forming a somewhat 1-sided spike; each flower is 3–5 cm long, dark blue with reddish-violet spots on the inside, erect and narrowly campanulate. There are 5 narrow pointed petal lobes, between which is a wide triangular 'tooth'. Leaves are ovoid-lanceolate, opposite, alternate, often in one plane when the stem is arching over.
Habitat and distribution
Cultivated in Britain. Damp meadows and woods on calcareous soils.
Additional information
Similar: Marsh Gentian (*G. pneumonanthe*) (see left), flowers usually solitary at stem end; green spotted stripes inside.

Stemless Gentian
Gentiana acaulis

Spring Gentian
Gentiana verna

Gentian family *Gentianaceae*
Flowering time: May–Aug.
Height of growth: 5–10 cm
Dicotyledonous; Perennial

Identification marks
Single, erect flower at stem end, 3–6 cm long. Peduncle very short. Only 1–2 pairs of stem leaves on the stalk, often no stem leaves at all. Remaining leaves in a basal tuft or rosette forming clumps.
Habitat and distribution
Cultivated in Britain. Mountains of Europe.
Additional information
2 groups within the species *G. acaulis*. *G. acaulis*, leaves approx. 8 cm long; flowers with conspicuous olive green longitudinal stripe inside. On acid soils. *G. clusii*, leaves approx. 2.5 cm long, rarely up to 6 cm. No olive green stripe on flower inside. On limy soils.

Gentian family *Gentianaceae*
Flowering time: April–Aug.
Height of growth: 3–15 cm
Dicotyledonous; Perennial

Identification marks
Single or, more rarely, 2–3 terminal flowers on erect stems each bearing 1–3 leaf pairs. Flowers deep blue, 2.5–3 cm long. Basal leaves in a tuft or a rosette, 1–3 cm long and approx. 1/2 as wide; blunt.
Habitat and distribution
Mountain meadows, open moorland, usually on limestone formations. Rare, but often in large numbers where it occurs.
Additional information
Does not tolerate fertilizers, which explains the reduction in its numbers over recent years. In the Alps several similar species grow which are not easily differentiated. They usually have smaller leaves.

Phacelia tanacetifolia

Common Comfrey
Symphytum officinale

Waterleaf family *Hydrophyllaceae*
Flowering time: June–Oct.
Height of growth: 15–50 cm
Dicotyledonous; Annual

Identification marks
Numerous flowers in compact cymes forming panicle-like inflorescence; lavender-coloured; campanulate to funnel-shaped; 4–6 mm across; the 5 stamens conspicuous – approx. twice as long as corolla, extending far beyond it. Stem erect, often branched, covered with coarse hairs. Leaves alternate, simple or bipinnately lobed.
Habitat and distribution
Waste ground; occasional.
Additional information
Originates in California. Introduced into Europe (including Britain) as an ornamental and as a bee plant.

Borage family *Boraginaceae*
Flowering time: May–June
Height of growth: 30–120 cm
Dicotyledonous; Perennial

Identification marks
Entire plant covered with rough hairs. Leaves rather narrowly ovoid, distinctly decurrent. Flowers small, campanulate, drooping, in scorpioidal cymes.
Habitat and distribution
On damp to wet ground, always rich in nutrients. In wet meadows, on river banks and in ditches. Found throughout Britain but less so in the north.

Soft Lungwort
Pulmonaria mollis

Pulmonaria obscura

Borage family *Boraginaceae*
Flowering time: April–May
Height of growth: 10–30 cm
Dicotyledonous; Perennial

Identification marks
Several flowers in cymes from the upper leaf axils. Flowers cowslip-like, with violet corollas which become lilac-coloured on fading. Basal leaves are narrower close to the stem; up to 45 cm long or more and approx. 1/3 of that in width. Leaves in the middle of the stem shorter. Whole plant softly hairy.
Habitat and distribution
Deciduous woodland on calcareous soil in central and south-eastern Europe. Scattered and local.
Additional information
Similar: *P. angustifolia*, basal leaves narrower towards the stem. Flowers bright blue. Local.

Borage family *Boraginaceae*
Flowering time: March–April
Height of growth: 15–40 cm
Dicotyledonous; Perennial

Identification marks
Several flowers in cymes from the upper leaf axils. Flowers cowslip-like, with red blossoms becoming violet then blue on fading. Basal leaves ovate-cordate, not or only very faintly spotted.
Habitat and distribution
Not British. Woodland, on calcareous formations. Scattered and local in northern Europe. Abundant in localities where it occurs.
Additional information
Similar: Lungwort (*P. officinalis*), peduncles of axillary inflorescences shorter than respective bract. Spotted leaves. Not British. Alps, Alpine foreland. Rare.

Field Forget-me-not
Myosotis arvensis

Water Forget-me-not
Myosotis scorpioides

Borage family *Boraginaceae*
Flowering time: May–Aug.
Height of growth: 10–30 cm
Dicotyledonous; Annual – Biennial

Identification marks
Numerous flowers in a relatively dense inflorescence each 3–4 mm across. Peduncles stand erect. During flowering they are 1–2 mm long. They spread as the fruit develops. Stem of fruit is 2–3 times as long as the calyx (which distinguishes it from small-flowered forms of Wood Forget-me-not). Basal leaves roundish ovate, petiolate, in rosettes, greyish-green. Stem leaves oblong-lanceolate, sessile.
Habitat and distribution
Roadsides, cultivated places and sandy dunes.
Additional information
Similar: Wood Forget-me-not (*M. sylvatica*), flowers 5–7 mm across, rarely smaller.

Borage family *Boraginaceae*
Flowering time: May–Oct.
Height of growth: 15–40 cm
Dicotyledonous; Perennial

Identification marks
10–20 flowers in a relatively lax raceme, flowers 4–10 mm across. Calyx has adpressed hairs (unlike Wood Forget-me-not where some individual hairs stand out on the calyx). Stem angular. Leaves elongate-lanceolate, sessile, hairy.
Habitat and distribution
Wet places, streamsides.
Additional information
There are closely related forms classed as species by some authorities. *M. caespitosa* stem round or indistinctly angular, branched from below the middle; scattered. *M. secunda* with fruiting pedicels 3–5 times the length of the calyx. Commonest in hilly districts.

Wood Forget-me-not
Myosotis sylvatica

Purple Gromwell
Buglossoides purpurocaeruleum

Borage family *Boraginaceae*
Flowering time: May–July
Height of growth: 15–50 cm
Dicotyledonous; Perennial

Identification marks
10–25 flowers in a lax raceme; each 5–10 mm across. Calyx has many individual hairs which stand out; occasionally there are only a few. Leaves in a rosette. Stem leaves becoming narrower, twice as long as they are wide.

Habitat and distribution
Woodland. Locally abundant.

Additional information
Sometimes confused with garden escapes which are often hybrids with *M. alpestris*, an alpine species from Europe.

Borage family *Boraginaceae*
Flowering time: April–June
Height of growth: 15–50 cm
Dicotyledonous; Perennial

Identification marks
Flowers in terminal cluster; buds brownish-red, young flowers reddish, then turning blue. Flowers 1–1.5 cm across. Stem erect, unbranched. Leaves lanceolate, up to 8 cm long, 1.5–2 cm wide.

Habitat and distribution
Dry deciduous woodland and thickets. Grows best in calcareous soil rich in humus. Rather rare, usually occurs in small numbers.

Additional information
In Britain it is confined to southern counties. In Europe it extends to the Mediterranean and eastwards to Asia Minor.

Bugloss
Anchusa arvensis

Vervain
Verbena officinalis

Borage family *Boraginaceae*
Flowering time: May–Oct.
Height of growth: 15–45 cm
Dicotyledonous; Annual

Identification marks
Numerous flowers in several simple or branched axillary cymes or clustered at the end. Flowers light blue, 5–7 mm across approx. 1 cm long. Corolla tube S-shaped and facing upwards. Stem angular. Leaves undulate with stiff hairs.
Habitat and distribution
Weedy places in fields, sandy heaths and near the sea. Prefers loose sandy ground. Rare.
Additional information
Used to be classed as *Lycopsis arvensis*. Its numbers have been greatly reduced by chemical pesticides; can still be found in field margins.

Verbena family *Verbenaceae*
Flowering time: July–Oct.
Height of growth: 30–60 cm
Dicotyledonous; Annual – Perennial

Identification marks
Numerous small (3–5 mm long) reddish-violet or pale lilac flowers in a spike-like inflorescence; conspicuously squarrosely branched. Leaves deeply divided, upper ones less so.
Habitat and distribution
Roadsides and waste places. Nitrogen indicator. Local. In Britain, commonest in the south.
Additional information
Probably originates from the Mediterranean. Requires warmth.

Bittersweet
Solanum dulcamara

Common Cornsalad
Valerianella locusta

Nightshade family *Solanaceae*
Flowering time: June–Aug.
Height of growth: 30–300 cm
Dicotyledonous

Identification marks
Flowers in umbellate clusters in leaf axils and at the stem end. Flowers violet, approx. 1 cm across; with 5 lobes usually folded back. Stem becomes woody; is erect or ascending. Leaves long-ovate glabrous, often deeply lobed.

Habitat and distribution
Grows in damp soil rich in nutrients. Woodland, fen carr, shingle beaches and waste ground. Common.

Additional information
Contains a poisonous alkaloid. The fruits are bright red berries which, if eaten, lead to severe poisoning.

Valerian family *Valerianaceae*
Flowering time: April–May
Height of growth: 5–40 cm
Dicotyledonous; Biennial

Identification marks
Several, small, inconspicuous flowers · in terminal groups of small clustered inflorescences. Flowers pale lilac blue. Stem erect and much branched. Lower stem leaves spatulate, upper ones lanceolate.

Habitat and distribution
Weedy areas in root crop fields, hedgebanks and dunes. Locally frequent.

Additional information
Several similar species can only be accurately identified by means of distinguishing features on the fruit wall. Grown as salad plants in numerous different varieties. Available when other salad plants virtually non-existent.

Alpine Bellflower
Campanula alpina

Bearded Bellflower
Campanula barbata

Bellflower family *Campanulaceae*
Flowering time: July–Aug.
Height of growth: 5–15cm
Dicotyledonous; Annual – Perennial

Identification marks
Usually 2–8 flowers (rarely only 1 or 2) in a short cluster. Flowers 3–4 cm long, light bluish-violet. The style divided into 3. Between each sepal, a short recurved tooth. Stem erect, with loose woolly hairs. Basal leaves and lower stem leaves narrow spatulate, slightly crenate at the apex, becoming gradually narrower towards the stem, with loose woolly covering of hairs.
Habitat and distribution
Not British or mainland European. Needs moist soil low in lime. Eastern Alps and mountains of the Balkan peninsula.
Additional information
May be confused with small examples of the Bearded Bellflower.

Bellflower family *Campanulaceae*
Flowering time: June–Aug.
Height of growth: 10–30 cm
Dicotyledonous; Perennial

Identification marks
2–12 flowers in a substantial cluster often 10 cm long or longer. Initially it is erect, later drooping. Flowers 1.5–3 cm long, light blue or pale lilac, with hairs along the veins and along both sides of the lobes. Narrow basal leaves.
Habitat and distribution
Not British. Prefers somewhat stony moist ground low in lime. Scattered in the mountains of central Alps, Europe, also in Norway.
Additional information
White-flowered forms are not infrequent.

Harebell
Campanula rotundifolia

Spreading Bellflower
Campanula patula

Bellflower family *Campanulaceae*
Flowering time: June–Sept.
Height of growth: 5–50 cm
Dicotyledonous; Perennial

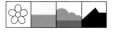

Identification marks
Up to 8 flowers in a lax panicle. Buds are erect but the flowers are pendent. Flowers 1.5–2 cm long, divided to no more than 1/3 their length. Leaves evenly distributed along the stem which has downy hairs at the base (check with magnifying glass). Stem leaves narrow lanceolate, margin entire, longer than 2 cm. Basal leaves have often withered by flowering time.
Habitat and distribution
Meadows, dry grassland. Very common.
Additional information
The Latin name indicates the first basal leaves which are orbicular in shape.

Bellflower family *Campanulaceae*
Flowering time: May–Sept.
Height of growth: 15–70 cm
Dicotyledonous; Annual – Biennial

Identification marks
Few flowers in freely branched panicles. Flowers erect on 2–5 cm, slender stalks; lobes conspicuously spread out, divided to approx. 1/2 the flower length. Erect branched stem has short hairs at the base.
Habitat and distribution
Hedgebanks, light woodland. Local.
Additional information
Easy to identify from other species because flowers are well spread out. Within the species there are virtually no varying forms.

Nettle-leaved Bellflower
Campanula trachelium

Creeping Bellflower
Campanula rapunculoides

Bellflower family *Campanulaceae*
Flowering time: May–Sept.
Height of growth: 50–100 cm
Dicotyledonous; Perennial

Identification marks
Flowers in a semi-erect cluster; 3.5–4.5 cm long; petal lobes have conspicuous hairs. Sepals have stiff hairs. Stem erect, sharply angled, hairy. Upper leaves sessile, lower stem leaves deeply heart-shaped with long stem. Petioles never winged.
Habitat and distribution
Semi-shade or shade and loamy soil, woodland. Scattered but widespread.
Additional information
Similar: Large Bellflower (*C. latifolia*), flowers 4–5 cm. Sepals glabrous. Stem bluntly angled. Petioles conspicuously winged. Woodland especially in the hilly areas of northern England and Scotland. Local.

Bellflower family *Campanulaceae*
Flowering time: June–Aug.
Height of growth: 30–60 cm
Dicotyledonous; Perennial

Identification marks
Flowers in a 1-sided raceme containing many flowers. They are nodding, 2–3 cm long; glabrous or have sparse long hairs along lobe edges; light violet. Stem round or slightly obtuse-angled. Stem leaves heart-shaped ovate or ovate-elongate. Basal leaves usually withered by flowering time.
Habitat and distribution
Fields and waste grassland, a weed in many old gardens.
Additional information
Creeping Bellflower has been introduced into Britain from Europe, where it grows as far east as Asia Minor and the Caucasus.

Peach-leaved Bellflower
Campanula persicifolia

Clustered Bellflower
Campanula glomerata

Bellflower family *Campanulaceae*
Flowering time: June–July
Height of growth: 70–120 cm
Dicotyledonous; Perennial

Identification marks
Few flowers in a semi-erect and 1-sided inflorescence. Flowers 2.5–4 cm across; usually equally long. No hairs on flower or calyx. Stem simple. Linear stem leaves are at most 1 cm wide, lower ones normally have small sharp saw-like teeth.

Habitat and distribution
Needs loamy woodland soil rich in nutrients. Scattered. Introduced to Britain from Europe or Asia where it is native. Established in several places but the colonies always start as garden escapes.

Bellflower family *Campanulaceae*
Flowering time: May–Sept.
Height of growth: 15–70 cm
Dicotyledonous; Perennial

Identification marks
Flowers stalkless in terminal heads, sometimes with a few axillary branches. They are virtually erect; 1.5–2.5 cm long, bluish-violet. Lower leaves rounded or heart-shaped. Entire plant covered with soft hairs.

Habitat and distribution
Grassy places on calcareous soils, occasionally in woods and on cliffs. Locally common.

Additional information
Similar: *C. cervicaria*, flowers 1–2 cm long, light bluish-violet. Style conspicuously longer than the flower. Lower leaves narrower towards the stem, never rounded. Leaves and stem covered with prickly stiff hairs. Not British. Thickets and meadows in Europe. Very rare.

Venus' Looking-glass
Legousia speculum-veneris

Phyteuma nigrum

Bellflower family *Campanulaceae*
Flowering time: June–Aug.
Height of growth: 10–20 cm
Dicotyledonous; Annual

Identification marks
Few flowers in a loose panicle; 1.5–2 cm across. Flower dark violet inside, somewhat lighter outside. Branched stem usually prostrate or turning upwards, more rarely erect; glabrous. Upper leaves lanceolate, sessile; lower ones lanceolate; petiolate.
Habitat and distribution
Not British. Needs calcareous soil rich in nutrients and a mild warm climate. Commonest in southern Europe.
Additional information
Similar: *L. hybrida*, flowers at the end of the stem clustered as on spike, otherwise racemose, only 0.8–1.5 cm across, purplish-red and lilac. Arable fields. Rare.

Bellflower family *Campanulaceae*
Flowering time: May–July
Height of growth: 20–70 cm
Dicotyledonous; Perennial

Identification marks
Flowers in a dense conical spike which is curved before opening, dark violet-blue. Bracts at base of spike usually shorter than spike. Basal leaves twice as long as wide.
Habitat and distribution
Not British. Needs loamy soil rich in nutrients and rather low in lime. Woodland, mountain meadows of the higher hilly regions of Europe.
Additional information
Similar: Blue-spiked Rampion (*P. betonicifolium*), flowers almost erect before opening; basal leaves approx. 3 times as long as wide; woodland; central Alps; scattered. Dark Rampion (*P. ovatum*), flowers bent upwards before opening; Alps; scattered.

Round-headed Rampion
Phyteuma orbiculare

Sheep's Bit
Jasione montana

Bellflower family *Campanulaceae*
Flowering time: May–Sept.
Height of growth: 10–50 cm
Dicotyledonous; Perennial

Identification marks
10–30 flowers in a globular head 1–2 cm across. Flowers bluish-violet, bent conspicuously inwards before opening. Basal leaves elongate-ovate, petiolate. Stem leaves elongate.

Habitat and distribution
Locally abundant in chalk grassland in the south.

Additional information
Abundant in central Europe, the Rampions reach their northerly limit in Britain.

Bellflower family *Campanulaceae*
Flowering time: June–Aug.
Height of growth: 10–50 cm
Dicotyledonous; Biennial

Identification marks
Numerous flowers in a globular head 1.2–2.5 cm across. Individual blossoms approx. 1 cm long, light bluish-violet. Occasionally, white flowers are found. Stem branched. Leaves lanceolate-ovate, glabrous or with stiff hairs, usually conspicuously undulate along the edges and often bluntly toothed.

Habitat and distribution
Sandy turf, rough grassland, heaths and cliffs. Locally abundant throughout Britain and Europe except for Mediterranean regions. In Britain it is commonest in Cornwall and in the Shetland Isles but can be found occasionally in most districts.

Lesser Grape Hyacinth
Muscari botryoides

Muscari racemosum

Lily family *Liliaceae*
Flowering time: April–May
Height of growth: 10–25 cm
Monocotyledonous; Perennial

Identification marks
Numerous flowers in a dense racemose inflorescence up to 2–5 cm long. Flowers approx. the same width as length; sky blue, drooping with white fringe around the apex. 2–3 basal leaves, almost flat.
Habitat and distribution
Not British. Mountain meadows, light open woodland on calcareous loamy soil. Virtually only higher hilly areas of mainland Europe and Alps.
Additional information
Similar: Tassel Hyacinth (*M. comosum*), flower cluster over 10 cm long when open; at the tip of the spike a crop of sterile blossoms face upwards. Gardens and waste ground.

Lily family *Liliaceae*
Flowering time: April
Height of growth: 10–40 cm
Monocotyledonous; Perennial

Identification marks
Numerous flowers in dense racemose spikes. Flower approx. twice as long as wide; sky blue, drooping, with white fringe around the mouth. Peduncle leafless. 4–6 basal leaves grooved, appearing in autumn, approx. 3 mm wide.
Habitat and distribution
Sunny, sandy turf. Rare.
Additional information
Similar: *M. neglectum*, flowers approx. 2½ times as long as they are wide. Inflorescence 2–4 cm long. Basal leaves approx. 5 mm wide, always longer than the stem. Frequent throughout Europe from the Mediterranean to southern Britain, but not in the north. Only native to sandy areas of East Anglian brickland.

Alpine Squill
Scilla bifolia

Iris sibirica

Lily family *Liliaceae*
Flowering time: March–April
Height of growth: 10–20 cm
Monocotyledonous; Perennial

Identification marks
2–8 flowers erect in a lax raceme; blue with a violet-coloured flush. Stem has no leaves. Usually 2 basal leaves, approx. 1 cm across, usually as long as the stem; hood-shaped at the apex.

Habitat and distribution
Not British. Occurs throughout mainland Europe in light woodland and somewhat moist soil rich in lime. Commonest in the Mediterranean region.

Additional information
Scilla verna, Spring Squill, which flowers in April and May, and *S. autumnalis*, Autumn Squill, flowering in August and September; native to coastal areas of southern England.

Iris family *Iridaceae*
Flowering time: May–July
Height of growth: 30–90 cm
Monocotyledonous; Perennial

Identification marks
Flowers large, blue or bluish-violet, 6 petals: 3 outer petals wide, 3 inner ones narrow. Deciduous leaves approx. 5 mm wide, shorter than stem, in 3-rowed arrangement overlapping at the base.

Habitat and distribution
Not British. Light woodland, wet meadows on ground rich in lime, and temporarily moist. Mainland Europe; never abundant.

Additional information
This plant does not survive either fertilizer or being cut down, which is why it dies out on grassy areas which are improved for grazing.

Spring Crocus
Crocus vernus

Purple Crocus
Crocus vernus ssp. *vernus*

Iris family *Iridaceae*
Flowering time: March–April
Height of growth: 5–15 cm
Monocotyledonous; Perennial

Identification marks
Leaves grass-like, with white central stripe, appearing shortly after the flowers. Flowers have narrow petals, at least 4 times longer than wide (2–3 cm long). Virtually no hairs in the flower throat.

Habitat and distribution
Native to Italy and the Balkan countries. In Britain it has become naturalized in meadows and pastures. Local in England and a few places in Scotland and Wales.

Additional information
The violet-flowering forms rarer than the white.

Iris family *Iridaceae*
Flowering time: March–April
Height of growth: 10–30 cm
Monocotyledonous; Perennial

Identification marks
Petals 2.5–4 cm long, 8–15 mm wide. Stigma longer than stamens. Virtually no hairs in the flower throat. Flowers deep violet. Leaves grasslike, with white central stripe, appearing shortly after flowers.

Habitat and distribution
In Britain only in gardens and occasionally naturalized.

Additional information
Originates around the Mediterranean. Brought to central Europe in the Middle Ages as an ornamental plant and has become naturalized in certain areas.

Pasque Flower
Pulsatilla vulgaris

Buttercup family *Ranunculaceae*
Flowering time: March–May
Height of growth: 5–40 cm
Dicotyledonous; Perennial

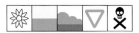

Identification marks
1 flower on each short peduncle; flower may reach 2–4 cm in length, lilac to dark violet. Petals have external covering of hairs. Funnel-shaped involucral bract below the flower, also densely covered with woolly hairs. Leaves dissected into many lobes, usually less than 5 mm wide.

Habitat and distribution
Needs calcareous soil and short grassland. Local in south-eastern England.

Additional information
A number of colour forms grown in gardens; pure white, red and occasionally pink flowers occur.

Hepatica
Hepatica nobilis

Buttercup family *Ranunculaceae*
Flowering time: March–May
Height of growth: 8–25 cm
Dicotyledonous; Perennial

Identification marks
Several stems arise from a leaf rosette, each stem bearing 1 flower. Flowers 2–3 cm across, normally blue, light bluish-violet or more rarely reddish-violet. Flowers have 6–10 petals. Leaves evergreen, trilobed, margin entire, often marbled with silver veining.

Habitat and distribution
Native to mainland Europe where it likes loamy soil rich in mull. Scattered, usually in large numbers where it occurs.

Additional information
Occasionally listed under Anemones (*Anemone hepatica*).

Alpine Snowbell
Soldanella alpina

Mountain Aster
Aster amellus

Primrose family *Primulaceae*
Flowering time: April–June
Height of growth: 5–15 cm
Dicotyledonous; Perennial

Identification marks
2–3 flowers on a peduncle; usually pendent, or partially erect; campanulate to funnel-shaped, violet to blue. Flower divided to 1/2 its length in a fringe of segments, 1–1.5 cm long. Leaves rounded, 1–3 cm wide, margin entire.

Habitat and distribution
Not British. Alps; snowy valleys and thickets alongside streams. Scattered.

Additional information
Soldanellas cultivated in rock gardens in Britain.

Daisy family *Asteraceae (Compositae)*
Flowering time: Aug.–Oct.
Height of growth: 15–50 cm
Dicotyledonous; Perennial

Identification marks
5–15 capitula in branched inflorescence; each capitulum 2–3 cm across. Individual flowers lilac or bluish-violet. 20–40 ray florets surround yellow disc florets. Bracts of capitula spread out. Stem erect, leaves lanceolate, hairy on the underside like the stem.

Habitat and distribution
Not British. In Europe found in light woodland, thickets, especially in the higher hilly areas where the sub-soil is limy. Rare.

Additional information
Similar: Sea Aster (*A. tripolium*), 26–80 capitula in 1 inflorescence. Bracts or capitula pressed close together. Plant glabrous. Only found on salty ground.

Cornflower
Centaurea cyanus

Mountain Knapweed
Centaurea montana

Daisy family *Asteraceae (Compositae)*
Flowering time: July–Oct.
Height of growth: 30–90 cm
Dicotyledonous; Annual

Identification marks
Capitula solitary; stem usually branched; flowers spread out, 3–5 cm across. Disc florets red purple, ray florets bright blue, and large. Stem erect. Leaves not decurrent, alternate, rarely exceeding 5 mm in width.

Habitat and distribution
Weedy places in cornfields, more rarely on waste ground. Rare, almost extinct.

Additional information
Cornflower is now absent from whole areas of the countryside, because of the use of chemical herbicides. Some forms still planted in the garden, often pink as well as blue.

Daisy family *Asteraceae (Compositae)*
Flowering time: May–Oct.
Height of growth: 30–60 cm
Dicotyledonous; Perennial

Identification marks
Flower head 4–6.5 cm across at the end of each unbranched stem. Outer florets larger, deep blue; inner florets blue/red/violet. The involucre of the head is made up of bracts with blackish-brown edges with 5–9 fringe segments on either side. Leaf margin entire, decurrent, fluffy.

Habitat and distribution
Not British. Native to mainland Europe. Needs calcareous loamy soil. Light woodland; mountain meadows. Higher hilly areas, Alps; scattered.

Additional information
Of the 400 species of *Centaurea* only *C. scabiosa* and *C. nigra* are frequently seen in Britain.

Creeping Thistle
Cirsium arvense

Chicory
Cichorium intybus

Daisy family *Asteraceae (Compositae)*
Flowering time: July–Sept.
Height of growth: 60–150 cm
Dicotyledonous; Perennial

Identification marks
Individual flowers in branched clusters. Each head 1–1.5 cm wide, with disc florets; usually a strong shade of lilac. Stem freely branched with mainly non-flowering branches. Leaves spiny, not decurrent, coarsely pinnate, usually with undulate margin.

Habitat and distribution
Weedy places in fields, gardens and waste places. Very common.

Additional information
The Creeping Thistle very variable; very dry habitats – conspicuously hairy; shady areas and on damp subsoil – usually quite glabrous. White-flowered plants occur.

Daisy family *Cichoriaceae (Compositae)*
Flowering time: July–Aug.
Height of growth: 30–130 cm
Dicotyledonous; Perennial

Identification marks
Compound inflorescence: numerous heads in a cluster on upper 2/3 of stem. Heads, 4–7 cm across, containing only ray florets; bright blue, occasionally pink or white. Stem branched. Lower leaves coarsely toothed, upper ones undivided, amplexicaul.

Habitat and distribution
Roadsides, dry grassland. Probably native in England and Wales but frequently a garden escape.

Additional information
Grown in 2 cultivated forms: one has rape-like roots (ssp. *sativa*) and provides the raw material for succory; the other form has a full leaf rosette (ssp. *foliosum*) and is grown as the salad plant.

Blue Lettuce
Lactuca perennis

Alpine Sow-thistle
Cicerbita alpina

Daisy family *Cichoriaceae (Compositae)*
Flowering time: May–June
Height of growth: 30–60 cm
Dicotyledonous; Perennial

Identification marks
Several heads 3.5–4.5 cm across, in a paniculate, almost umbellate inflorescence. The heads contain 14–18 ray florets, normally pure blue, perhaps also bluish-violet or reddish-violet. Leaf margins entire, more often pinnate.

Habitat and distribution
Not British. Native to Europe where it is found in calcareous, stony, loamy ground, dry turf and thickets.

Additional information
Flower heads close close in the afternoon and in dull weather. Rare nowadays because of the disappearance of suitable habitats. Once common locally and used for making salads.

Daisy family *Cichoriaceae (Compositae)*
Flowering time: July–Sept.
Height of growth: 50–200 cm
Dicotyledonous; Perennial

Identification marks
Numerous heads in spike-like clusters at the end of the stems. These heads contain only bluish-violet ray florets. The stems of the inflorescence covered with reddish glandular hairs. Leaves divided into coarse toothed lobes, the lower more so than the upper. The terminal lobe spear-shaped.

Habitat and distribution.
Needs damp soil rich in nitrogen and low in lime. Found on meadows in the hilly regions in the Alps. In Britain known only on alpine rocks in eastern Scotland.

Additional information
The exact status of this plant in Britain is doubtful, many botanists believing it an introduction.

Limodore
Limodorum abortivum

Forking Larkspur
Consolida regalis

Orchid family *Orchidaceae*
Flowering time: May–July
Height of growth: 10–50 cm
Monocotyledonous; Perennial

Identification marks
Entire plant leafless, usually tinged with blue or violet. Flowers 1.5–2.5 cm across; yellow, with violet tinge; lip usually violet with yellowish shading. Not all flowers open fully.
Habitat and distribution
Not British. Loamy or loess soil with sufficient humus. Prefers semi-shade. Thickets. Native to Europe. It grows in the Upper Rhine area as far north as the Eiger Mountains in Germany and in the warmest Alpine valleys.
Additional information
The main area of distribution is the Mediterranean. Limodore does not flower every year even in suitable habitats.

Buttercup family *Ranunculaceae*
Flowering time: May–Sept.
Height of growth: 10–50 cm
Dicotyledonous; Annual

Identification marks
3–7 flowers in a sparse forking cluster; flowers 1.5–2.5 cm wide, the spur up to 2.5 cm long. Usually dark blue, occasionally dark violet. Leaves deeply divided, double trilobed.
Habitat and distribution
Not British (only found in Britain as a casual). Weedy places in cornfields or on dry waste ground. Mainland Europe.
Additional information
Forking Larkspur has disappeared from many parts because of chemical herbicides. Similar: *C. ajacis*, with usually more than 7 flowers in a dense cluster, blue and sometimes pink and white. Leaves never double trilobed but double pinnate, the pinnae themselves dissected. Rare in a few places.

Monk's-hood
Aconitum napellus

Lupinus polyphyllus

Buttercup family *Ranunculaceae*
Flowering time: June–July
Height of growth: 60–150 cm
Dicotyledonous; Perennial

Identification marks
Flowers in terminal racemes; uppermost petal of each flower forms a wide helmet. Stem erect. Leaves petiolate and palmately lobed, 5–7 segments. Leaf sections are divided into narrow lobes.

Habitat and distribution
Needs damp soil. In mainland Europe it is found in mountain meadows and in the highest hilly regions, and in the Alps. In Britain it is by streams, usually under light tree cover.

Additional information
Many variations of this species; frequently subdivided into subspecies.

Pea family *Fabaceae (Leguminosae)*
Flowering time: June–Sept.
Height of growth: 50–150 cm
Dicotyledonous; Perennial

Identification marks
Numerous flowers in an erect cluster, 15–60 cm long. Flowers 1.2–1.5 cm long, usually blue, more rarely violet or even red or white. Leaves palmately divided into 9–17 leaflets, up to 15 cm long 3 cm wide.

Habitat and distribution
Not British. Rare in Europe but usually where it does occur it is in conspicuous abundance.

Additional information
The roots contain nitrogen-forming bacteria which helps to improve the soil. For this reason it is planted in some regions as an agricultural crop.

Lucerne
Medicago sativa

Oxytropis montana

Pea family *Fabaceae (Leguminosae)*
Flowering time: June–Sept.
Height of growth: 20–80 cm
Dicotyledonous; Perennial

Identification marks
Numerous flowers in small, head-like clusters, 2–3 cm long and not quite as wide; lilac, violet or reddish-violet. Stem branched. Leaves trilobed with ovate lanceolate leaflets. Terminal leaflet has distinctly longer petiole, up to 3 cm.

Habitat and distribution
Roadsides, grassland, forest rides. Naturalized in light soils.

Additional information
Lucerne is widely grown as a fodder plant; escapes from cultivation not uncommon. Where it grows near the native Sickle Medick (*M. falcata*) it gives rise to hybrids with green and even black flowers.

Pea family *Fabaceae (Leguminosae)*
Flowering time: July–Aug.
Height of growth: 5–15 cm
Dicotyledonous; Perennial

Identification marks
Flowers drooping, initially in dense clusters, the stalks lengthening later so they become lax; bluish-violet. Keel pointed. Calyx teeth 1/4–1/3 as long as calyx tube. Leaves have petiole tinged with red, pinnate, with 25–41 leaflets, almost glabrous.

Habitat and distribution
Not British. Native to the Alps.

Additional information
Similar: *O. campestris*, dull yellow, and *O. halleri*, pale purple; both native to upland areas of Scotland.

Lathyrus liniifolius

Spring Pea
Lathyrus vernus

Pea family *Fabaceae (Leguminosae)*
Flowering time: April–June
Height of growth: 15–40 cm
Dicotyledonous; Perennial

Identification marks
3–6 flowers in lax axillary cluster, up to 7 cm long. Flowers initially red, then dirty blue, 11–22 cm long. Stamens tubular, tube has straight edge. 4–6 leaflets per leaf, normally distinctly bluish-green on the underside. Short point at end of leaf. Stem narrowly but conspicuously winged.
Habitat and distribution
Not British. In Europe found in woodland, heaths, mountain meadows.
Additional information
Occasionally confused with the Spring Pea (right). Its winged stem is easily recognizable, however.

Pea family *Fabaceae (Leguminosae)*
Flowering time: April–June
Height of growth: 20–60 cm
Dicotyledonous; Perennial

Identification marks
2–7 flowers in lax axillary cluster, up to 6 cm long. Flowers initially red, then dirty blue, 1.5–2 cm long. Stamens tubular, tube has straight edge. 4–6 leaflets per leaf, each 1/2 as wide as long. Short point at leaf end. Stem quadrangular and quite definitely not winged.
Habitat and distribution
Not British. In Europe found in woodland, preferably deciduous. Prefers calcareous soil. Scattered, but locally common.
Additional information
Occasionally confused with *Lathyrus liniifolius* (left). Stem not winged, therefore easy to recognize.

Smooth Tare
Vicia tetrasperma

Bush Vetch
Vivia sepium

Pea family *Fabaceae (Leguminosae)*
Flowering time: June–July
Height of growth: 20–60 cm
Dicotyledonous; Annual

Identification marks
1–3 flowers in a cluster with long stem. Flowers pale violet to lilac, approx. 5 mm long. Fruit usually has 4 seeds, rarely 5. Pinnate leaves have 6–10 leaflets and a tendril at the leaf apex.

Habitat and distribution
Weedy areas of cornfields, grassland and waste places. Scattered.

Additional information
Similar: Hairy Tare (*V. hirsuta*), usually only 1 flower, pale blue, almost white, 5 mm long; fruit has 2 seeds; 12–20 leaflets per leaf; weedy areas; scattered. *V. tenuissima*, 1–3 flowers, pale blue, 8 mm long; fruit 5–6 seeds; 4–8 leaflets per leaf; weedy places; rare in southern England.

Pea family *Fabaceae (Leguminosae)*
Flowering time: May–Aug.
Height of growth: 30–60 cm
Dicotyledonous; Perennial

Identification marks
2–6 flowers in very short-stemmed clusters, spread out or drooping in axils of upper leaves. Stamens tubular, tube has crooked edge. Leaves pinnate, 8–16 leaflets, with pinnate tendril at the end. Stem climbs with leaf tendrils.

Habitat and distribution
Grassland, hedgebanks and light woodland. Common.

Additional information
Some botanists recognize many variations within this species. The groups are distinguished on the basis of the variations in hair covering on the calyx and the differing width of the leaflets. The differences are, however, not always easy to spot and not always constant.

Tufted Vetch
Vicia cracca

Fine-leaved Vetch
Vicia tenuifolia

Pea family *Fabaceae (Leguminosae)*
Flowering time: June–Aug.
Height of growth: 30–150 cm
Dicotyledonous; Perennial

Identification marks
20–40 flowers in a long-stemmed cluster. Flowers around 1 cm long, bluish-violet. Stamens tubular, tube has crooked edge. 12–20 leaflets. Instead of terminal leaf, a tendril usually with 2 lateral tendrils. Leaflets almost glabrous on the uppersides. Peduncle approx. 3/4 of leaf length out of whose axil it grows.
Habitat and distribution
Grassland, light woodland. Scattered.
Additional information
Similar: *V. dasycarpa*, only 5–15 flowers in a cluster. Peduncle approx. 3/4 of leaf length out of whose axil it is growing. 12–20 leaflets with conspicuous hairs.

Pea family *Fabaceae (Leguminosae)*
Flowering time: June–July
Height of growth: 50–100 cm
Dicotyledonous; Perennial

Identification marks
20–40 flowers in a long-stemmed cluster. Peduncle as long or longer than leaf out of whose axil it grows. Leaves 18–28 leaflets and a tendril at the end. Leaflets almost glabrous on uppersides, sparse clinging hairs on undersides.
Habitat and distribution
A casual from Europe found on waste ground in a few localities. Rare.
Additional information
Similar: *V. villosa*, 10–30 flowers in a long-stemmed cluster. Peduncle approx. as long or longer than the leaf out of whose axil it is growing. 10–20 leaflets with projecting hairs. An occasional casual.

Dwarf Milkwort
Polygala amarella

Common Milkwort
Polygala vulgaris

Milkwort family *Polygalaceae*
Flowering time: May–June
Height of growth: 5–15 cm
Dicotyledonous; Perennial

Identification marks
10–40 flowers in a cluster; usually blue, more rarely reddish, each 2–4 mm long. Stem prostrate at the base, then ascending to erect. Leaves form a basal rosette, otherwise alternate, leaves taste bitter if chewed.

Habitat and distribution
Needs moist soil rich in lime. Found in chalk grassland in southern England. Rare.

Additional information
Similar: Bitter Milkwort (*P. amara*), flowers larger (3–7 mm), inflorescence more lax. Moist hillside pastures on limestone in northern England. These species resemble each other closely, classed together under the name *P. amara*.

Milkwort family *Polygalaceae*
Flowering time: May–Aug.
Height of growth: 10–50 cm
Dicotyledonous; Perennial

Identification marks
5–30 flowers in a cluster; usually blue, lilac, more rarely red; each approx. 8 mm long. Bracts of the inflorescence 2 mm and therefore not projecting beyond flowers before blossoming. Stem erect or ascending. Leaves alternate, never in a rosette.

Habitat and distribution
Meadows. Found also on heaths and dunes throughout Britain.

Additional information
In Britain can only be confused with *P. serpyllifolia*; alternate leaves; at least the lowest leaves opposite.

Heath Milkwort
Polygala serpyllifolia

Bog Violet
Viola palustris

Milkwort family *Polygalaceae*
Flowering time: May–July
Height of growth: 5–20 cm
Dicotyledonous; Perennial

Identification marks
3–10 flowers in a cluster; usually blue, occasionally almost white, approx. 5 mm long. Bracts approx. 1 mm long. Stem prostrate at base, then curved upwards and erect. Lower leaves opposite, but never in a rosette.

Habitat and distribution
Grassland which is low in nutrients and lime. Widespread throughout Britain.

Additional information
Since the plant needs both a humid climate as well as soil low in lime it is found most frequently in north-western Europe, not extending as far south as *P. amara*.

Violet family *Violaceae*
Flowering time: May–July
Height of growth: 8–15 cm
Dicotyledonous; Perennial

Identification marks
2–6 flowers in a basal rosette. Peduncles solitary, growing out of the axils of the rosette leaves; approx. as long as the petioles and glabrous. Sepals blunt, glabrous. Flowers pale lilac, with dark brownish-violet veins. Flowers about 1.5 cm across.

Habitat and distribution
Needs wet soil low in nutrients. Common except in the drier parts of eastern England.

Additional information
Similar: *V. stagnina* (Fen Violet), which occurs only in calcareous fens in eastern England and in Ireland. It has bluish-white flowers and a short spur.

Heath Dog Violet
Viola canina

Early Dog Violet
Viola reichenbachiana

Violet family *Violaceae*
Flowering time: April–June
Height of growth: 5–15 cm
Dicotyledonous; Perennial

Identification marks
No rosette. All leaves borne along the stem, usually considerably longer than 2 cm and often longer than they are wide. Petal spur approx. 1.5 cm long. Flower barely longer than it is wide; bluish-violet.

Habitat and distribution
Needs soil low in lime. Prefers sandy ground. Frequent throughout Britain.

Additional information
As well as the ssp. *canina* there are at least 2 more, one of which is rare in eastern England: ssp. *montana*, petal distinctly longer than it is wide. Damp areas on heathland; rare.

Violet family *Violaceae*
Flowering time: April–May
Height of growth: 3–20 cm
Dicotyledonous; Perennial

Identification marks
Basal leaves present, stem has foliage. Single axillary flowers; petal spur approx. 2 cm long or longer. Stem prostrate or ascending. Leaves crenate. Stipules have long fringes. Plant glabrous.

Habitat and distribution
Woodland where ground is rich in humus.

Additional information
The Early Dog Violet is close to *V. riviniana*, Common Dog Violet. They are most easily distinguished by their spurs. *V. reichenbachiana* can be identified mainly by its slender violet-coloured spur. *V. riviniana* has a blunter spur which is white in colour.

Sweet Violet
Viola odorata

Hairy Violet
Viola hirta

Violet family *Violaceae*
Flowering time: Feb.–April
Height of growth: 3–10 cm
Dicotyledonous; Perennial

Identification marks
All leaves are basal, the stems being leafless. Flowers dark violet, scented. Petal with a spur approx. 15 mm long, the same colour as the petals. Leaves wide ovate to reniform, crenate. There is often a second flowering in autumn.

Habitat and distribution
Dry thickets, hedgerows. Likes ground rich in nitrogen. Scattered.

Additional information
Sweet Violet is frequently planted, but many of its habitats close to houses may be the remains of wild populations. In cultivation there are forms with different flower colours (white, also found in the wild, pink and pale yellow).

Violet family *Violaceae*
Flowering time: April–May
Height of growth: 5–15 cm
Dicotyledonous; Perennial

Identification marks
All leaves basal. Peduncles grow singly or in small numbers from the leaf axils. Leaves usually more triangular in shape, always with at least a sparse covering of hairs. Petal spur approx. 17 mm long. Stipules (press the rosette apart to see them) are wide lanceolate, glabrous, with few fringes.

Habitat and distribution
Needs calcareous soil. Meadows, woodland. Local.

Additional information
Because of its preference for calcareous grassland, this species is often found on prehistoric banks and earthworks in southern England.

Viper's Bugloss
Echium vulgare

Ground Ivy
Glechoma hederacea

Borage family *Boraginaceae*
Flowering time: June–Sept.
Height of growth: 30–120 cm
Dicotyledonous; Perennial

Identification marks
Flowers single or in small groups in leaf axils on the upper 1/2 of stem. Flowers are red in bud, turning blue when open. Stem erect. Leaves lanceolate. Entire plant covered with bristles which are conspicuously thickened at their base.

Habitat and distribution
Roadsides, waste places. Once a serious weed of arable fields. Common, especially in the south.

Additional information
Occasionally occurring as a casual: Pale Bugloss (*E. italicum*), flowers light violet, lilac or white, approx. 1 cm long. Calyx only covered with dense bristles. Probably originates in the eastern Mediterranean.

Mint family *Lamiaceae (Labiatae)*
Flowering time: May–June
Height of growth: 15–60 cm
Dicotyledonous; Perennial

Identification marks
2–3 flowers in leaf axils of the upper 1/2 of the stem; 1–2 cm long, bluish-violet. The upper lip is flat and thus inconspicuous. Stem creeping, ascending or erect. Leaves petiolate, reniform or cordiform, crenate.

Habitat and distribution
Woodland, meadows, waste places. Common.

Additional information
This species is very varied. The size of the plants is usually determined by environmental conditions. It is remarkably tolerant of soil moisture, growing on dry road banks and also in pathways in marshland and woods.

Bugle
Ajuga reptans

Ajuga genevensis

Mint family *Lamiaceae (Labiatae)*
Flowering time: May–June
Height of growth: 15–30 cm
Dicotyledonous; Perennial

Identification marks
Flowers without upper lip and with trilobed lower lip, 6–12 in false whorls in upper leaf axils, spike-like at the end of the stem. Bracts undivided. Basal leaves in a rosette, spathulate, slightly crenate. Plant spreads by runners along the ground.

Habitat and distribution
Woodland, damp meadows, on loamy somewhat moist soil rich in nutrients. Very common in suitable habitats.

Additional information
Considerable variation in size and hair covering, depending on habitat. Occasionally also plants with pink or white flowers. In this case the chromoplast formation, which is controlled by several genes, has not been functioning properly.

Mint family *Lamiaceae (Labiatae)*
Flowering time: April–June
Height of growth: 5–30 cm
Dicotyledonous; Perennial

Identification marks
Flowers without upper lip and with trilobed lower lip, 6–12 in false whorls in upper leaf axils, spike-like at the end of the stem. Bracts deeply cleft into 3 lobes, rarely just toothed. Plant has no runners.

Habitat and distribution
Grows on dry ground. Occasionally naturalized but not wild in Britain.

Additional information
Similar: Pyramidal Bugle (*A. pyramidalis*), no runners. Leaves grow close to the stem. Basal leaves up to 10 cm long and 5 cm wide. Stem leaves smaller, margin entire or slightly crenate, often tinged with violet. Found in rock crevices in higher hilly areas in the north. Rare.

Skullcap
Scutellaria galericulata

Lesser Skullcap
Scutellaria minor

Mint family *Lamiaceae (Labiatae)*
Flowering time: June–Sept.
Height of growth: 10–50 cm
Dicotyledonous; Perennial

Identification marks
In the upper third of the stem there are 1–4 pairs of flowers in the leaf axils forming a 1-sided spike. Flowers bluish-violet in colour, 10–20 mm long. Lower lip usually lighter. Stem usually erect.

Habitat and distribution
Needs wet ground. Reed beds, wet woodland areas, water meadows. Common in suitable habitats.

Additional information
The commonest of the 13 species which occur in Europe, being found in all mainland parts of the continent except the extreme north and south.

Mint family *Lamiaceae (Labiatae)*
Flowering time: July–Oct.
Height of growth: 10–15 cm
Dicotyledonous; Perennial

Identification marks
1–3 pairs of flowers in the leaf axils of the upper third of the stem making a 1-sided spike. Flowers light pinkish-purple, occasionally with a touch of red, only 0.5–0.8 cm long. Stem ascending or erect.

Habitat and distribution
Needs wet acid ground, usually found on wet heaths. Throughout Britain except the far north-east.

Additional information
Similar: Spear-leaved Skullcap (*S. hastifolia*), in the upper fifth of the stem there are 2–5 pairs of flowers borne in a terminal raceme. Flowers blue, 1.8–2.5 cm long. Middle leaves have conspicuous spear-shaped lobes. Very rare, one site where it is a certain introduction.

Self-heal
Prunella vulgaris

Large Self-heal
Prunella grandiflora

Mint family *Lamiaceae (Labiatae)*
Flowering time: May–Oct.
Height of growth: 10–20 cm
Dicotyledonous; Perennial

Identification marks
Flowers in a rounded cluster at the stem end; 1–1.5 cm long; upper lip helmet-shaped and flat. The calyx approx. 2/3 the flower length. Directly below inflorescence is a pair of opposite leaves.

Habitat and distribution
Needs loamy nitrogenous soil. Meadows and woodland. Very common.

Additional information
A plant of wide tolerances, being found in damp woods, in all sorts of grassland. In gardens it is a common weed of damp lawns.

Mint family *Lamiaceae (Labiatae)*
Flowering time: June–Aug.
Height of growth: 5–25 cm
Dicotyledonous; Perennial

Identification marks
Flowers dark violet, in rounded clusters at stem ends, 2–2.5 cm long. Upper lip helmet-shaped; calyx approx. 1/2 flower length. First pair of stem leaves a conspicuous distance from the inflorescence. Leaf margin crenate or entire.

Habitat and distribution
Not native in Britain. Needs calcareous soil which is warm in the summer. Dry turf.

Additional information
Becoming more common as an ornamental rock plant; occasionally also with white or red flowers.

Meadow Sage
Salvia pratensis

Whorled Clary
Salvia verticillata

Mint family *Lamiaceae (Labiatae)*
Flowering time: May–July
Height of growth: 20–60 cm
Dicotyledonous; Perennial

Identification marks
4–8 flowers in whorl-like inflorescences in the stem in terminal spikes, sometimes branched; hairy. Flowers usually bluish-violet, more rarely light blue, pink or white, 2–2.5 cm long, with very inflated upper lip. Leaves in a loose rosette with few stem leaves; leaves wrinkled, coarsely toothed-crenate.
Habitat and distribution
Needs calcareous soil rich in nutrients. Rare in chalk grassland in southern England.
Additional information
Similar: *S. verbenaca*, flowers only 0.8–1.5 cm long. Lowest bract in the inflorescence always longer than the respective petals. In Britain grows only in Guernsey.

Mint family *Lamiaceae (Labiatae)*
Flowering time: June–Sept.
Height of growth: 30–60 cm
Dicotyledonous; Perennial

Identification marks
12–24 flowers in whorls in the upper 1/3 o the stem in a loose spike; flowers 1–1.5 cm long; dark violet. Stem erect and hairy. Basa leaves usually withered by flowering time Stem leaves ovate, heart-shaped at the base and usually with projecting lobes.
Habitat and distribution
A rare plant occasionally found in waste places.
Additional information
Probably originates in south-east Europe From there it has gradually spread northwards. It has established itsel particularly in warmer regions.

Basil Thyme
Acinos arvensis

Alpine Bartsia
Bartsia alpina

Mint family *Lamiaceae (Labiatae)*
Flowering time: June–Aug.
Height of growth: 10–30 cm
Dicotyledonous; Biennial – Perennial

Figwort family *Scrophulariaceae*
Flowering time: June–Aug.
Height of growth: 5–20 cm
Dicotyledonous; Perennial

Identification marks
2–3 flowers in whorls above one another on the upper part of the stem and also from leaf axis. Flowers 0.7–1 cm long; bluish-violet. Stem prostrate, ascending or erect, usually branched from the base and sparsely hairy. Leaves opposite, with short petioles, barely 2 cm long, ovoid.

Habitat and distribution
Dry turf, scree along path verges. Likes warmth. Scattered.

Additional information
Basil Thyme is very varied. Over the last few decades it has been classified under various names: *Calamintha acinos, Satureja acinos, Satureja calamintha.*

Identification marks
Single flowers in upper leaf axils; 1.5–2.5 cm long; very dark violet; frequently indistinguishable from the bracts which are also tinged with dark violet. Upper lip of the flower inflated, the lower lip flat and trilobed. Leaves ovoidly crenate, opposite, hairy and bluntly toothed.

Habitat and distribution
Needs damp ground rich in nutrients. Moorland in hilly regions of northern England and Scotland.

Additional information
Semi-parasite; uses sucker roots to draw nutritive salts and water from the plants whose roots it taps.

Alpine Toadflax
Linaria alpina

Ivy-leaved Toadflax
Cymbalaria muralis

Figwort family *Scrophulariaceae*
Flowering time: June–July
Height of growth: 5–10 cm
Dicotyledonous; Perennial

Identification marks
2–8 flowers in a dense terminal cluster; reddish or bluish-violet, except throat which is almost always reddish-orange. Very rarely the flowers are entirely yellow. Flowers 1–1.5 cm long; upper lip bifid. Stem prostrate and only curved upwards at the tips. 3–4 leaves in a whorl; lanceolate margin entire.

Additional information
Not British. Alps and Alpine foothills, on scree and stony turf. Scattered, but occurs locally in strikingly large numbers.

Figwort family *Scrophulariaceae*
Flowering time: June–Aug.
Height of growth: 30–60 cm
Dicotyledonous; Perennial

Identification marks
Single axillary flowers on long stems. Flower approx. 7 mm long, light violet with yellow palate. Spur blunt, short and only about 1/ flower length. The often purplish stem is ver thin and grows prostrate or hanging roote along walls or rocks. Leaves widely hear shaped, with 5–7 shallow palmate lobe glabrous, undersides usually reddish.

Habitat and distribution
Common on walls. It has established itself i Britain from gardens but is not found in wil habitats.

Additional information
Originates from the northern Mediterranean

Small Toadflax
Chaenorhinum minus

Common Butterwort
Pinguicula vulgaris

Figwort family *Scrophulariaceae*
Flowering time: June–Sept.
Height of growth: 5–20 cm
Dicotyledonous; Annual

Identification marks
Single axillary flowers on relatively long
peduncles on the upper 1/2 of the stems.
Flowers approx. 7 mm long, violet, the lower
lip almost white. Spur short (2–3 mm),
straight and somewhat pointed. Ripe
capsules often tinged with red and from a
distance resemble flowers. Usually few open
flowers on the plant. Stem erect, leaves
narrow-lanceolate, lower ones opposite.
Habitat and distribution
Arable fields, railway ballast, roadsides and
waste places. Scattered, especially in
southern England.
Additional information
Also known under the name of *Linaria minor*;
regarded as a typical railway weed.
Originates from the Mediterranean.

Butterwort family *Lentibulariaceae*
Flowering time: May–June
Height of growth: 5–15 cm
Dicotyledonous; Perennial

Identification marks
Flowers solitary, bluish-violet, somewhat
lighter in the throat, with spur. Upper lip only
slightly curved upwards. Lobes of the lower
lip do not overlap each other or only a little.
Leaves in basal rosette, yellowish, sticky,
curled up around the edges.
Habitat and distribution
Bogs and seepage areas between rocks. Rare,
often in decreasing numbers.
Additional information
Similar: *P. grandiflora*, much larger with
flowers to 20 mm; native to western Ireland.
P. lusitanica, pale lilac and white flowers;
found in the west of Britain.

Globularia punctata

Small Scabious
Scabiosa columbaria

Globularia family *Globulariaceae*
Flowering time: May–July
Height of growth: 5–30 cm
Dicotyledonous; Perennial

Identification marks
Numerous flowers in a rounded head; not surrounded by bracts; may reach 1–1.5 cm across. Flowers approx. 7 mm long and bluish-violet. Basal leaves in a loose rosette, round to ovoid, narrower towards the stem. Stem leaves much smaller, ovate, sessile.
Habitat and distribution
Dry chalky turf. Cultivated in Britain.
Additional information
Similar: *G. nudicaulis*, capitula 1.5–2.5 cm across. Alps. Not in Britain though both are grown in gardens.

Scabious family *Dipsacaceae*
Flowering time: June–Oct.
Height of growth: 30–60 cm
Dicotyledonous; Annual – Perennial

Identification marks
Flowers in a head surrounded by bracts. Marginal florets larger than the inner ones. Flowers lilac to violet. Basal leaves and lowest pair of stem leaves usually undivided. Stem leaves pinnate, sparsely hairy.
Habitat and distribution
Dry chalky grassland, rocks, cliffs. Frequent in suitable habitats.
Additional information
Similar: Shiny Scabious (*S. lucida*), flowers reddish or bluish-violet. Basal leaves and lowest pair of stem leaves have short hairs along the margin and the veins only, otherwise glabrous and somewhat glossy. Only in the Alps.

Field Scabious
Knautia arvensis

Knautia dipsacifolia

Scabious family *Dipsacaceae*
Flowering time: June–Sept.
Height of growth: 30–70 cm
Dicotyledonous; Perennial

Identification marks
Flowers in a head surrounded by bracts. Marginal florets larger than the inner ones. Flowers lilac, violet or reddish-violet. Stem beneath the flower head covered with conspicuously projecting hairs. Leaves opposite, greyish-green. At least the upper stem leaves pinnately divided.

Habitat and distribution
Meadows, roadsides, weedy communities on fallow land. Very common.

Additional information
Very varied. Might be confused with others in the genus. However, it is the most common of all similar plants in Britain and over most of Europe.

Scabious family *Dipsacaceae*
Flowering time: June–Sept.
Height of growth: 30–100 cm
Dicotyledonous; Perennial

Identification marks
Flowers in a head surrounded by bracts. Bracts almost as long as the flowers or longer. Marginal florets larger than inner ones. Flowers lilac or light bluish-violet. Stem almost completely covered with hairs, especially lower down. All leaves undivided, opposite, usually conspicuously hairy.

Habitat and distribution
Not British. Mountain forests in Europe which are not too dry, Alpine pastures.

Additional information
In Europe there are similar species which are hard to distinguish from one another. In Britain only *K. arvensis* (left) is similar.

Greater Reedmace
Typha latifolia

Branched Bur-reed
Sparganium erectum

Reedmace family *Typhaceae*
Flowering time: June–Aug.
Height of growth: 90–250 cm
Monocotyledonous; Perennial

Identification marks
Male spadix almost directly above the female which is more or less the same length. The female part of the spike is velvety and blackish-brown. Leaves are grass-like, slightly twisted, 1–2 cm wide. Plant forms creeping shoots and usually grows in dense clumps.

Habitat and distribution
Reedbeds in stagnant or slow-flowing waters, ditches, ponds, canals. Frequent in suitable habitats.

Additional information
Similar: Lesser Reedmace (*T. angustifolia*), male and female have a spadix 3–8 cm length of stem between them. Leaves only 0.5–1 cm wide. Reedbeds. Widely distributed.

Bur-reed family *Sparganiaceae*
Flowering time: June–Aug.
Height of growth: 50–150 cm
Monocotyledonous; Perennial

Identification marks
Leaves grass-like, stiff. Stem branched. Male and female flowers in separate round heads at end of branches, upper ones being male, lower ones female; of bur-like appearance when fruit is ripe.

Habitat and distribution
Frequently found in reedbeds of stagnant or slow-flowing waters, in ditches and marshes. Likes nutrients. Common in almost all areas.

Additional information
Similar: rare species which do not however have a branched stem: Unbranched Bur-reed (*S. emersum*), Floating Bur-reed (*S. angustifolium*) with floating leaves, Least Bur-reed (*S. minimum*) with 2–5 capitula.

Broad-leaved Pondweed
Potamogeton natans

Common Duckweed
Lemna minor

Pondweed family *Potamogetonaceae*
Flowering time: June–Aug.
Height of growth: 50–150 cm
Monocotyledonous; Perennial

Identification marks
Flowers in a spike up to 8 cm long held several centimetres above water by 10 cm long stem. Individual flowers inconspicuous. Leaves ovate floating on the water, twice as long as they are wide, borne on long stems.

Habitat and distribution
In shallow water (under 1 m in depth) in ponds, ditches and slow rivers. Common usually in large numbers in those localities where it occurs.

Additional information
Similar: *P. polygonifolius*, submerged leaves present all year round, lanceolate, transparent. Floating leaves barely longer than they are wide. Bog pools with acid water. Common.

Duckweed family *Lemnaceae*
Flowering time: June–Aug.
Height of growth: 50–150 cm
Monocotyledonous; Perennial

Identification marks
Plant composed of leaf-like structures (thalli) which float. Thalli grow 1.5–2 times as long as they are wide, 2–6mm in length; flat (not inflated on the underside). their roots hang free in the water.

Habitat and distribution
Stagnant and slow-flowing waters.

Additional information
Similar: Ivy Duckweed (*L. trisulca*); leaf-like structures lanceolate, pointed, in opposite pairs, many together in a group, 4–10 mm long. Stagnant waters. Scattered. Gibbous Duckweed (*L. gibba*): leaf-like structures 2–5 mm long, inflated onthe underside. Stagnant waters. Rare.

Lords and Ladies
Arum maculatum

Sweet Flag
Acorus calamus

Arum family *Araceae*
Flowering time: April–May
Height of growth: 30–45 cm
Monocotyledonous; Perennial

Identification marks
Greenish-white spathe encloses the spadix, male flowers above the female. The upper part of the spadix stalked and club-shaped, club emitting a smell of carrion. This serves as a fly-trap for the purposes of pollination. Basal deciduous leaves sagittate; sometimes black-spotted.

Habitat and distribution
England and Wales, rarer in Scotland. On loose soil rich in nutrients in deciduous and mixed woodland, in thickets, also along hedgerows. Prefers habitats where the soil is loamy and rich in mull; likes warm situations.

Additional information
No possibility of confusion. The club can vary from white to violet. Other common names: Cuckoo-pint, Jack-in-the-pulpit.

Arum family *Araceae*
Flowering time: April–June
Height of growth: 15–50 cm
Monocotyledonous; Perennial

Identification marks
Inconspicuous flowers in a lateral spadix, 4–10 cm long. Below spadix is a bract which can be 2–10 times longer than the spadix. Stem triangular in section; leaves linear, reed-like, margin entire, often wrinkled on one side.

Habitat and distribution
Reedbeds of stagnant or slow-flowing waters. Rare. Found locally throughout Britain usually in shallow water.

Additional information
Cultivated since the 16th century for strewing on floors and since then has escaped from gardens and become naturalized.

Hop
Humulus lupulus

Asarabacca
Asarum europaeum

Hemp family *Cannabaceae*
Flowering time: July–Aug.
Height of growth: 1–7 m
Dicotyledonous; Perennial

Identification marks
Climbing plant. Plant dioecious; male and female flowers on different plants. Female flowers in yellowish-green cone-like catkins, 2–5 cm long. Leaves opposite, palmately lobed, 3–7 segments (apart from uppermost leaves of female plant which are not divided).
Habitat and distribution
Lowland forests, damp thickets. Likes warm situations. Widely naturalized.
Additional information
Hops are extensively cultivated, probably grown since the 8th century. They contain a bitter substance which is important for the flavour of beer.

Birthwort family *Aristolochiaceae*
Flowering time: April–May
Height of growth: 5–10 cm
Dicotyledonous; Perennial

Identification marks
Single axillary flowers, lying on the ground or only slightly raised; occasionally also hidden under foliage, 1.5–2 cm long, brownish-green on the outside, reddish-brown on the inside. Leaves petiolate, reniform, glossy, winter-green, basal.
Habitat and distribution
Needs calcareous soil containing humus, predominantly in deciduous, or more rarely mixed, woodland or coniferous forests. Possibly native in a few places in southern England. Rare.
Additional information
Contains a burning ethereal oil which is at least slightly poisonous.

Common Nettle
Urtica dioica

Wood Dock
Rumex sanguineus

Nettle family *Urticaceae*
Flowering time: June–Oct.
Height of growth: 60–150 cm
Dicotyledonous; Perennial

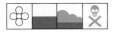

Identification marks
Male and female flowers on separate plants. Flower panicles long, pendulous. Plant has stinging hairs. Leaves opposite, longer than their petioles, usually over 5 cm, and 2–3 cm longer than they are wide.
Habitat and distribution
Waste places, roadsides, woodland. Nitrogen and wetness indicator. Very common.
Additional information
Similar: Small Nettle (*Urtica urens*), male and female flowers on the same inflorescence, usually more female than male. Leaves normally less than 5 cm in length and max. 1 1/2 times as long as they are wide. Waste places and cultivated ground. Common in south-east England. A non-stinging form rarely occurs.

Dock family *Polygonaceae*
Flowering time: June–Sept.
Height of growth: 30–70 cm
Dicotyledonous; Perennial

Identification marks
Flowers inconspicuous, in whorls on the upper 1/2 of the stem. Inflorescence has no leaves or at most only a lanceolate bract on each of lower flower whorls. Petals round-ovoid; very raised swellings. Often entire inflorescence tinged red, usually just the petals have a reddish tinge.
Habitat and distribution
Woodland paths, light woodland areas, watersides. Always in ground at least occasionally damp. More common in southern Britain.
Additional information
Similar: Clustered Dock (*R. conglomeratus*), inflorescence has lanceolate bracts almost to apex of each of the whorls. River banks and light woodland areas.

Curled Dock
Rumex crispus

Broad-leaved Dock
Rumex obtusifolius

Dock family *Polygonaceae*
Flowering time: July–Aug.
Height of growth: 30–100 cm
Dicotyledonous; Perennial

Dock family *Polygonaceae*
Flowering time: June–Aug.
Height of growth: 50–120 cm
Dicotyledonous; Perennial

Identification marks
Flowers carried above the leaves in loose, sometimes branched spikes made up of whorls. Individual flowers inconspicuous. Leaves tongue-shaped and pronouncedly curled at the edges. Lowest leaves up to 30 cm long.

Habitat and distribution
Weedy communities in waste ground and roadsides. Likes nitrogen. Common.

Additional information
Similar: Great Water Dock (*R. hydrolapathum*), leaves not curled at the edges, broadly lanceolate and much larger than those of *R. crispus*. Lowest leaves 30–80 cm long. Reedbeds of stagnant or slow-flowing water, also on waste ground which is temporarily flooded. Widespread.

Identification marks
Branched panicles made up of dense flower whorls. Lateral branches of inflorescence erect, not branched, brownish-red when plant bears fruit. Basal leaves large, with heart-shaped base and slightly undulating edges. There are bracts on lower 1/2 of the inflorescence.

Habitat and distribution
Weedy communities on waste ground and roadsides.

Additional information
Similar: Golden Dock (*R. maritimus*), basal leaves usually less than 20 cm long, never have heart-shaped base but narrow down towards the stem. Inflorescence has bracts right to apex, yellow when bearing fruit, never reddish-brown. Local. In muddy places by ponds, reservoirs and lakes.

Monk's Rhubarb
Rumex alpinus

Glasswort
Salicornia europaea

Dock family *Polygonaceae*
Flowering time: July–Aug.
Height of growth: 50–200 cm
Dicotyledonous; Perennial

Identification marks
Inflorescence freely branched. Lateral branches erect and curved towards the main stem, making inflorescence noticeably dense. Often rust-brown as petals of the inconspicuous flowers are frequently tinged red. Basal leaves up to 60 cm long, ovate to round, almost as on Butterburs. Stem has longitudinal grooves.
Habitat and distribution
Native to mainland Europe but introduced into Britain where it occurs locally.
Additional information
Monk's Rhubarb contains a great deal of oxalic acid and is therefore avoided by most grazing cattle.

Goosefoot family *Chenopodiaceae*
Flowering time: Aug.–Nov.
Height of growth: 5–45 cm
Dicotyledonous; Annual – Biennial

Identification marks
Flowers inconspicuous, hidden behind small scales at the ends of the branches which are thickened and club-shaped. Stems fleshy, articulated, freely branched especially at the base. Branches curved upwards. Occasionally prostrate, more usually ascending or erect and green, greenish-yellow or dirt red.
Habitat and distribution
Silted mudflats. Often planted along flat heavily silted littorals to help reclaim the land. Frequent along coast and usually in large numbers.
Additional information
Glasswort is one of the native plants which needs salt. Many closely related species are recognized.

Fat Hen
Chenopodium album

Good King Henry
Chenopodium bonus-henricus

Goosefoot family *Chenopodiaceae*
Flowering time: July–Sept.
Height of growth: 20–150 cm
Dicotyledonous; Annual

Identification marks
Flowers inconspicuous, in small cymes which form a freely branched inflorescence. Clustered spikelets in the leaf axils, or compressed to form a compound inflorescence. Flower clusters mealy. Stem angular. Leaves lanceolate-rhomboid, also mealy.
Habitat and distribution
Prefers ground rich in nitrogen. Wasteland, rubbish dumps, gardens, manure heaps. Very common.
Additional information
Very varied in appearance. The species can be divided into several races but it is not easy to differentiate clearly between them.

Goosefoot family *Chenopodiaceae*
Flowering time: May–Aug.
Height of growth: 10–60 cm
Dicotyledonous; Perennial

Identification marks
Flowers inconspicuous, in greenish.yellow dense cymes forming narrow cylindrical terminal panicle. All leaves sagittate, margin entire, covered with mealy powder on the undersides at least when plant is young. Plant dull green. Leaves often have undulate margins. Stem also slightly mealy beneath the inflorescence.
Habitat and distribution
Frequently found in weedy communities close to farms and cottages and on waste ground. Frequent. Not truly native.
Additional information
Used to be eaten as a vegetable prepared like spinach, and is still cultivated though not so frequently as in the past.

Spear-leaved Orache
Atriplex hastata

Common Orache
Atriplex patula

Goosefoot family *Chenopodiaceae*
Flowering time: July–Oct.
Height of growth: 30–100 cm
Dicotyledonous; Annual

Identification marks
Flowers inconspicuous, in greenish clusters arranged in spike-like inflorescences borne in the leaf axils or at the end of the branches. The female flowers have 2 spear-shaped bracts conspicuous when the plant is fruiting. Male flowers are on the same plant (monoecious). Leaves opposite or alternate, margin entire or toothed, glabrous or mealy. Lower leaves triangular to spear-shaped.

Habitat and distribution
Weedy communities in fields, along paths, on waste ground and along the coast.

Additional information
Similar: Grass-leaved Orache (*A. littoralis*), all leaves narrowly lanceolate, the lower ones also never hastate. Frequent along coastal banks.

Goosefoot family *Chenopodiaceae*
Flowering time: July–Oct.
Height of growth: 30–100 cm
Dicotyledonous; Annual

Identification marks
Flowers inconspicuous, in spike-like inflorescences tinged with green or red, in the leaf axils and at the ends of the branches. The female flowers have two triangular bracts 3–6 mm long which are conspicuous in fruit. Male flowers are on the same plant (monoecious), usually not in large numbers. All leaves alternate, rhomboid in shape, often hastate at the base.

Habitat and distribution
Weedy communities on wasteland, along paths and on fields. Common.

Additional information
The only species most commonly found inland. *A. glabriuscula*, Babington's Orache and *A. laciniata*, Frosted Orache, are found on the shore line. Both are silvery mealy.

Common Amaranthus
Amaranthus retroflexus

White Pigweed
Amaranthus albus

Cockscomb family *Amaranthaceae*
Flowering time: July–Sept.
Height of growth: 10–80 cm
Dicotyledonous; Annual

Identification marks
Numerous inconspicuous flowers in dense spikes in axils of upper leaves and at ends of stems. Flower bracts narrow into a spiny thorn. Petals have prominent central rib (check with magnifying glass). Stem erect, covered with downy hairs. Leaves rhomboid, up to 12 cm long, more than twice as long as they are wide.
Habitat and distribution
Weedy communities. Rare.
Additional information
A. hybridus, stem glabrous, flower bracts twice as long as the leaves which are pointed and where the central rib does not project (check with magnifying glass). Waste ground. Not in Britain.

Cockscomb family *Amaranthaceae*
Flowering time: Aug.–Oct.
Height of growth: 20–50 cm
Dicotyledonous; Annual

Identification marks
Few inconspicuous flowers in simple spikes in axils of upper leaves. No spike at end of stem. Spikes light green. Flower bracts twice as long as petals, gradually narrowing down to form a spiny thorn. Stem prostrate to erect, light green, glabrous. Leaves ovate only approx. 2 cm long, with cartilaginous undulating margin, leafy to top of stem.
Habitat and distribution
Weedy communities on rubbish tips, railway tracks, and waste ground. Rare.
Additional information
No species of *Amaranthus* is native to Britain; occurs only as a casual.

Parsley Piert
Aphanes arvensis

Lady's Mantle
Alchemilla vulgaris

Rose family *Rosaceae*
Flowering time: May–Oct.
Height of growth: 5–10 cm
Dicotyledonous; Annual

Identification marks
10–20 flowers in dense clusters borne opposite the leaves. Stipules below flower heads all partly fused together with the petiole to form a cup. Flower approx. 1.5–2 mm long and made up only of calyx which is spread out. Leaves have 3–5 lobes. Plant is prostrate or slightly ascending.

Habitat and distribution
Weedy areas especially in cornfields and dry sandy soils. Common.

Additional information
Similar: *A. microcarpa*, flowers not even 1 mm long. Sepals never spread out. Cornfields and on sandy soils. Less common.

Rose family *Rosaceae*
Flowering time: May–June
Height of growth: 20–70 cm
Dicotyledonous; Perennial

Identification marks
Flowers in a terminal panicle which is glabrous at the top. Flowers small and comprise only sepals. Rounded-reniform leaves have 7–11 lobes and are toothed. Between teeth are pores which secrete drops of water on damp nights.

Habitat and distribution
Meadows, woodland paths, roadsides. Common.

Additional information
Species is normally divided into many micro-species. The reason lies in the method of propagation of the plant. Ovaries are not pollinated, in fact fertile pollen is not normally formed. Instead, vegetative embryos grow out of embryo-sac.

Salad Burnet
Sanguisorba minor

Great Burnet
Sanguisorba officinalis

Rose family *Rosaceae*
Flowering time: May–June
Height of growth: 20–70 cm
Dicotyledonous; Perennial

Identification marks
Flowers in a tight globular head 1–1.5 cm across. Flowers green, sepals (petals absent) are tinged with red or brown along the edge. The styles are purple-red giving their colour to the flower. Leaves pinnate with 5–17 round and toothed leaflets. They are 10–20 cm long.
Habitat and distribution
Dry grassland, chalky banks. Likes warmth. Widespread and common except in Scotland.
Additional information
A good indicator plant of chalky soils usually found with Rock Rose and Dropwort.

Rose family *Rosaceae*
Flowering time: June–Sept.
Height of growth: 5–15 cm
Dicotyledonous; Perennial

Identification marks
Dense terminal heads of deep purplish-red or brownish-red flowers. Stem erect and branched in the upper part. Leaves large, pinnate. Leaflets petiolate, ovoid, margin crenate.
Habitat and distribution
Moist meadows. Grows on very damp, often peaty soil but also loamy. Does not form very large numbers. Common, locally, mainly in western parts of Britain.
Additional information
The inconspicuous flowers are generally bisexual, i.e. they have both stamens and ovaries. Occasionally, however, you may find unisexual flowers. The very light seeds of the Great Burnet are dispersed by the wind.

Petty Spurge
Euphorbia peplis

Sweet Spurge
Euphorbia dulcis

Spurge family *Euphorbiaceae*
Flowering time: June–Nov.
Height of growth: 5–35 cm
Dicotyledonous; Annual

Identification marks
Inflorescence umbellate usually with 3 main rays which are in turn branched. Glands in bracts are crescent-shaped (check with magnifying glass). Stem erect. Leaves on stem are alternate, soon falling. If present they are generally on a conspicuous petiole from the stem. Plant contains white milky juice.
Habitat and distribution
Weedy areas in gardens, root-crop fields, waste places, roadsides. Common.
Additional information
At first glance *E. helioscopa*, Sun Spurge, may be confused with this species. However, the former usually has 5 main branches in its umbellate inflorescence.

Spurge family *Euphorbiaceae*
Flowering time: May–June
Height of growth: 10–50 cm
Dicotyledonous; Perennial

Identification marks
Umbellate inflorescence with 3–5 main rays which are not further branched or only forked once. Glands in bract are oval; in young plants they are yellow, then green, and finally reddish-yellow to purplish-red. Stem erect. Leaves 2.5–6 cm long and 1–2 cm wide invert ovoid.
Habitat and distribution
Introduced and naturalized in a few places mainly in Scotland.
Additional information
Similar: *E. cyparissias*, Cypress Spurge, 9–1 rayed umbels; spreads with great rapidity in sandy soils; naturalized in southern England.

Dog's Mercury
Mercurialis perennis

Annual Mercury
Mercurialis annua

Spurge family *Euphorbiaceae*
Flowering time: April–May
Height of growth: 15–30 cm
Dicotyledonous; Perennial

Identification marks
Male and female flowers on separate plants. Dioecious. Male flowers in clusters on long-stalked spikes, female 1–3 on long stalks. Stem erect, with leaves in upper 1/2 only, unbranched. Leaves petiolate, ovate-lanceolate, dark green, 3 times as long as wide. Plant smells slightly unpleasant when rubbed.

Habitat and distribution
Woodland, hedgerows. Very common and often in abundance in its habitats.

Additional information
An indicator of ancient woodland, it is rarely found where woodland has not occurred in the past.

Spurge family *Euphorbiaceae*
Flowering time: June–Oct.
Height of growth: 25–50 cm
Dicotyledonous; Annual

Identification marks
Male and female flowers on separate plants. Dioecious. Up to 10 male flowers in clustered spikes in the leaf axils, female almost stalkless in the leaf axils. Stem erect, with leaves along its length. Leaves opposite. Stem has lateral branches and is quadrangular. Leaves petiolate, ovatelanceolate, crenate. Plant smells unpleasant when rubbed.

Habitat and distribution
Weedy areas. Scattered.

Additional information
Widespread throughout lowland Britain, often as a weed in parks and gardens. In some places plants of only one sex occur. On such occasions one plant spreads vegetatively.

Water-starwort
Callitriche palustris

Starwort family *Callitrichaceae*
Flowering time: June–Oct
Height of growth: 2–20 cm
Dicotyledonous; Perennial

Identification marks
Plant comprises a floating leaf rosette. Leaves up to 2 cm long, fairly narrow, tongue-shaped in appearance. Floating leaf rosette attached to a very thin, thread-like, submerged stem. Plant monoecious. Flowers very inconspicuous.

Habitat and distribution
Slow-flowing or stagnant water. Land forms are occasionally found on dried-up mud. Scattered.

Additional information
There are several very closely related species and subspecies characterized by fruit structure and also partly by fine details in leaf shape.

Mare's-tail
Hippuris vulgaris

Mare's-tail family *Hippuridaceae*
Flowering time: May–July
Height of growth: 10–150 cm
Dicotyledonous; Perennial

Identification marks
Flowers small, inconspicuous in axils of upper leaves, comprising only a stamen and ovary. Stem is thick, up to 1 cm across bearing 4–20 linear leaves in each whorl. Whorls fairly compressed. Leaves needle-like or tongue-shaped. Submerged leaves limp.

Habitat and distribution
Stagnant and slow-flowing water. Local. Usually occurs in large numbers.

Additional information
The Mare's-tail spreads by its runners and by shoots which break off and form roots.

Great Plantain
Plantago major

Ribwort Plantain
Plantago lanceolata

Plantain family *Plantaginaceae*
Flowering time: June–Oct.
Height of growth: 15–30 cm
Dicotyledonous; Perennial

Identification marks
Individual flowers inconspicuous, in a long spike on a stem which is shorter than the leaves. Stamens yellowish-white, pollen sacs reddish-violet. Leaves in a basal rosette ascending or erect at an angle.

Habitat and distribution
Weedy communities along paths, roadsides, waste places, meadows and pastures. Likes nitrogen, resistant to trampling. Very common.

Additional information
The growth point and flower spikes arise from below the leaves which protect them from injury.

Plantain family *Plantaginaceae*
Flowering time: May–Oct.
Height of growth: 5–60 cm
Dicotyledonous; Perennial

Identification marks
Individual flowers inconspicuous, carried in a short cylindrical spike. Stamens white, later brown. Petals barely 2 mm long; brown. Stem erect, grooved. Leaves in a rosette, at least partly erect, 3–10 times as long as wide, with 3–7 conspicuous veins, hence its common name.

Habitat and distribution
Meadows, pastures, roadsides, waste ground. Very common.

Additional information
Similar: *P. media*, Hoary Plaintain, finely hairy all over; cylindrical flower spikes have pinky-white anthers. Common on limy soils.

Pale Persicaria
Polygonum lapathifolium

Annual Knawel
Scleranthus annuus

Dock family *Polygonaceae*
Flowering time: June–Oct.
Height of growth: 25–75 cm
Dicotyledonous; Annual

Identification marks
Stem erect or ascending in bends, usually richly branched, with thickened nodes. Leaves oval, broadest over the basal 1/3, often with dark spots. Leaf sheath (ochra) membraneous, cornet-shaped, the uppermost with short cilia. Flower spikes at ends of branches, many-flowered, white or greenish or rarely pink.
Habitat and distribution
Weedy places in fields, on banks of rivers and ponds, less common along paths or in water. Likes moisture and plenty of nutrients. Common.
Additional information
Similar: Redshank (*P. persicaria*, p. 214), all leaf sheaths have fringe of cilia, often grow together in similar habitats. Flowers also white or red but usually pink.

Pink family *Caryophyllaceae*
Flowering time: May–Oct.
Height of growth: 8–20 cm
Dicotyledonous; Annual

Identification marks
Flowers inconspicuous, in dense clusters in leaf axils or at ends of branches. No petals, 5 sepals 2 mm long, 2–5 stamens barely 1 mm long. Sepals ovoid, pointed, with very narrow dry membraneous white margins. Stem erect. Leaves opposite or clustered, linear or awl-shaped.
Habitat and distribution
Weedy areas in fields and gardens. Prefers sandy ground which is at least slightly acid. Scattered.
Additional information
Divided into several subspecies. The differences between them lie in characteristics of fruit and sepals.

Stinking Hellebore
Helleborus foetidus

Green Hellebore
Helleborus viridus

Buttercup family *Ranunculaceae*
Flowering time: March–April
Height of growth: 30–50 cm
Dicotyledonous; Perennial

Identification marks
Paniculate pendulous inflorescence made up of several nodding, bell-shaped flowers. Sepals overlap one another, often with red margin at the apex. The stem is leafy to the base. Leaves are palmately lobed; becoming less dissected to the top of the stem where they merge into bracts. No actual basal leaves present. Lower stem leaves evergreen.

Habitat and distribution
Grows in calcareous soil. Dry woodland, more rarely scree. Scattered and probably native in southern England.

Additional information
In Europe a plant of western, more maritime regions.

Buttercup family *Ranunculaceae*
Flowering time: March–May
Height of growth: 30–50 cm
Dicotyledonous; Perennial

Identification marks
The 2 basal leaves are deciduous. 1–4 flowers somewhat bent over but not totally pendent. No petals. Sepals wide ovoid, spread out, overlapping at the edges. Leaves divided into 7–11 lobes, outer lobes further divided. Segments doubly serrate.

Habitat and distribution
Needs moist calcareous ground containing humus. Deciduous woodland. Local in England.

Additional information
Several Hellebores are widely cultivated but none have become naturalized, so there is unlikely to be any confusion over identification.

Water Avens
Geum rivale

Garden Angelica
Angelica archangelica

Rose family *Rosaceae*
Flowering time: May–Sept.
Height of growth: 20–60 cm
Dicotyledonous; Perennial

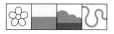

Identification marks
Several flowers in a lax cyme, drooping. Reddish-brown sepals conspicuous. Petals dull orange-pink. Leaves alternate, irregularly pinnate, upper ones trifoliate.

Habitat and distribution
Wet meadows, flat moors, ditches, wet woodland and rock ledges. Common in northern England, Wales and Scotland, local or rare in southern England.

Additional information
The individual plants within the species vary very greatly, but so far it has not been possible to clearly define any further divisions. In larger groups one often finds 'anomalies': the bracts form an integral part of the flowers, the flower does not droop, etc.

Umbellifer family *Apiaceae (Umbelliferae)*
Flowering time: June–July
Height of growth: 50–250 cm
Dicotyledonous; Biennial

Identification marks
Flowers in compound umbels. Main umbel has 20–30 rays (branches). Stem is floury round the umbel, round, hollow, branched at the top, green and as broad as a human arm at the base. Leaf sheaths inflated.

Habitat and distribution
Weedy communities at watersides, in reedbeds. Rare. Not native in Britain though locally naturalized.

Additional information
Similar: Giant Hogweed (*Heracleum mantegazzianum*): marginal florets in the umbels conspicuously enlarged. Leaves are coarsely trifoliate, often over 1 m long. Entire plant up to 3 m high and more. Ornamental plant, cultivated and also growing wild in meadows along streams. Both species can cause blisters when sap gets on skin later exposed to sun.

Deadly Nightshade
Atropa bella-donna

Moschatel
Adoxa moschatellina

Nightshade family *Solanaceae*
Flowering time: June–July
Height of growth: 5–150 cm
Dicotyledonous; Perennial

Moschatel family *Adoxaceae*
Flowering time: March–April
Height of growth: 5–10 cm
Dicotyledonous; Perennial

Identification marks
Single flowers in axils of upper leaves. They have a greenish-red tinge; lobes are deep brownish-red, brownish-violet or purplish-violet. Stem erect, leaves ovate, decurrent. Usually a large leaf grows next to a small leaf. Fruit a berry, large and black.

Habitat and distribution
Woodland and thickets on limy soils. Rare.

Additional information
Deadly Nightshade is deadly poisonous. It contains hyoscyamine and smaller quantities of atropine.

Identification marks
The flowers are in 5s in a small terminal head, 4 facing outwards, 1 upwards; usually yellowish-green and approx. 5 mm across. Topmost flower has 4 petals, the others have 5 petal lobes. Stem erect. Basal leaves have long petioles, doubly trifoliate.

Habitat and distribution
Needs ground rich in nutrients. Damp woodland, hedgerows and rock ledges. Widespread but nowhere common.

Additional information
Freely translated *adoxa* means 'not worth mentioning'. When Linnaeus used this description he wished to convey the fact that he was not concerned about the variable number of petals on the flower.

Rannoch Rush
Scheuchzeria palustris

Fritillary
Fritillaria meleagris

Arrow-brass family *Scheuchzeriaceae*
Flowering time: May–July
Height of growth: 10–30 cm
Monocotyledonous; Perennial

Identification marks
3–10 flowers in a lax raceme shorter than the leaves. Peduncles up to 1 cm long. The flowers themselves only approx. 5 mm long, yellowish-green. 6 stamens at least as long as the petals. Leaves rush-like, the basal sheath enlarged. Deciduous leaves from previous year form a circle of withered leaf sheaths at the end of the horizontal rootstock.

Habitat and distribution
Pools in sphagnum bogs. Very rare.

Additional information
Distributed across entire northern hemisphere where climate is moderate to cold. Seeds of the plant have been found in deposits from between the Ice Ages.

Lily family *Liliaceae*
Flowering time: April–May
Height of growth: 20–50 cm
Monocotyledonous; Perennial

Identification marks
Stem erect, with 3–6 very narrow deciduous leaves; these are grooved and blue-green. Flowers solitary (rarely in pairs) drooping, campanulate, up to 4 cm long and 2 cm wide, white and purple chequered like a chess board, rarely creamy.

Habitat and distribution
Very local, in low-lying water meadows rich in nutrients, often flooded in spring. S., E., and C. England. Often only planted.

Additional information
Often cultivated in gardens and then sometimes escaping. Formerly quite widespread but now disappearing.

Herb Paris
Paris quadrifolia

Alpine Leek
Allium victorialis

Lily family *Liliaceae*
Flowering time: May–June
Height of growth: 15–30 cm
Monocotyledonous; Perennial

Identification marks
Only 1 terminal flower; up to 5 cm across. Peduncle which carries the 4-petalled flower above the whorl of deciduous leaves is always leafless. Leaf whorl consists of 4 leaves. Fruit is a bluish-black berry about the size of a cherry.

Habitat and distribution
Needs shade and a somewhat damp soil rich in nutrients. Woodland. Widespread.

Additional information
Often individual plants in a larger number will have more leaves in the whorl (frequently 5, more rarely 6) or the number of petals varies. It is not yet clear how far this is determined by genetics or by environmental influences.

Lily family *Liliaceae*
Flowering time: July–Aug.
Height of growth: 30–60 cm
Monocotyledonous; Perennial

Identification marks
Globular inflorescence. Petals approx. 5 mm long, somewhat blunt. Before opening, inflorescence is often drooping, surrounded by a white membraneous bract. Stem rounded, 2–3 leaves in lower 1/2. Leaves elliptical-lanceolate, 2–3 cm wide, with short petioles.

Habitat and distribution
Not British. Alps. Needs moist soil containing nutrients. Grows in woods, thickets and stony places.

Additional information
Rootstock often Mandrake-like in shape. It has always been associated with mystical concepts, for example the possessor of the Mandrake is said to be invulnerable.

Broad-leaved Helleborine
Epipactis helleborine

Orchid family *Orchidaceae*
Flowering time: June–Aug.
Height of growth: 20–70 cm
Monocotyledonous; Perennial

Identification marks
15–30 flowers in a lax raceme. Lip does not have a spur; apex constricted laterally and therefore conspicuously bilobed. Petals noticeably campanulate, light green or slightly tinged with brown. Leaves ovate, 1.5–3 times as long as they are wide, always 2–3 times longer than the stem section between 2 leaves.

Habitat and distribution
Needs loamy soil. Woodland, hedgebanks. Scattered throughout Britain.

Additional information
Similar: *E. purpurata*, flower spread out, hardly campanulate in formation; lip without spur; inflorescence 15–25 cm long, dense, with many flowers. Entire plant tinged with violet. Limy soil. Woodland. Rare.

Common Twayblade
Listera ovata

Orchid family *Orchidaceae*
Flowering time: May–June
Height of growth: 20–65 cm
Monocotyledonous; Perennial

Identification marks
20–40 flowers in a long lax raceme. Flowers green. Labellum long, pendulous, bilobed without spur, up to 1 cm long, remaining petals considerably shorter and forming a helmet shape. Stem has only 2 opposite leaves which are leathery, wide ovate and up to 10 cm long.

Habitat and distribution
Woodland, damp grassland and dunes. Common and widespread.

Additional information
Similar: Lesser Twayblade (*L. cordata*), much smaller, never over 20 cm, flower green, labellum reddish. Inflorescence has only 5–10 flowers. Labellum 4–8 mm long, usually with pointed lobes. Only 2 opposite cordiform leaves. Coniferous woodland and peaty moors. Commonest in Scotland, not in south-east England.

Bird's-nest Orchid
Neottia nidus-avis

Frog Orchid
Coeloglossum viride

Orchid family *Orchidaceae*
Flowering time: June–July
Height of growth: 20–45 cm
Monocotyledonous; Perennial

Identification marks
Entire plant yellowish-brown. Tough stem with narrow-ovoid leaves. Raceme with many flowers, cylindrical. Flowers do not have a spur. Labellum bilobed, upper perianth-segments coming together to form a head.

Habitat and distribution
In deciduous and mixed woodland, especially in beechwoods. Likes loamy soils rich in lime and nutrients and absorbs organic material in the mull with the help of fungi. Throughout Britain but easily overlooked.

Additional information
The rootstock under the ground bears many tightly interwoven roots. Their nest-like appearance gave the plant its name.

Orchid family *Orchidaceae*
Flowering time: May–June
Height of growth: 10–30 cm
Monocotyledonous; Perennial

Identification marks
Inflorescence 3–10 cm long and relatively dense. Upper petals in a helmet formation approx. 5 mm across. Lip drooping, approx. 8 mm long, undivided, tongue-shaped, with 3 teeth only at the apex. Spur is short. Stem angular, with 2–5 alternate leaves.

Habitat and distribution
Grasslands and pastures especially on chalky soils, also on moist hillsides to 1,000 m in Scotland.

Additional information
Similar: Musk Orchid (*Herminium monorchis*), only 5–10 cm high, 10–30 small flowers which smell distinctly of honey. Lip is barely 5 mm long, trilobed. Stem usually has 2 narrow ovoid leaves. Limy grassland. Rare. Only in southern England.

Ophrys holosericea
Orchid family

Early Spider Orchid
Ophrys sphegodes

Orchid family *Orchidaceae*
Flowering time: June–July
Height of growth: 15–30 cm
Monocotyledonous; Perennial

Orchid family *Orchidaceae*
Flowering time: May–June
Height of growth: 15–30 cm
Monocotyledonous; Perennial

Identification marks
Few flowers, very lax in an almost 1-sided spike. Marginal petals spread out, white or red. Lip without spur, covered with velvety hairs, slightly convex and approx. as wide as long. Lip a mixture of purple, brown and red to deep brown, with yellow markings; at the top an upwardly-curved 'beak'. Leaves elongate-ovate, alternate.

Identification marks
Only 2–3 flowers (rarely as many as 8) in a lax and almost 1-sided spike. Marginal petals spread out, green or greenish-yellow. Lip without spur, covered with velvety hairs, pronouncedly convex, about as wide as long. Lip dark brown with blue markings which resemble an H or a horse-shoe. Leaves elongate-ovate, alternate.

Habitat and distribution
Not British. In Europe semi-dry turf, meadows with poor soil, glades in dry woodland. Needs a warm climate.

Habitat and distribution
Limy grassland, meadows with poor soil, glades in dry woodland. Needs a warm climate. Rare.

Additional information
When this plant appears together with other species of the same family, hybrids quite frequently occur.

Additional information
When the Early Spider Orchid appears together with other species of the same family, hybrids may occur.

Fly Orchid
Ophrys insectifera

Bee Orchid
Ophrys apifera

Orchid family *Orchidaceae*
Flowering time: May–July
Height of growth: 10–40 cm
Monocotyledonous; Perennial

Identification marks
2–15 flowers in a lax almost 1-sided spike. Marginal petals spread out, green to yellowish-green. Lip without spur, covered with velvety hairs, longer than wide, trilobed.

Habitat and distribution
Limy grassland and light areas in dry woodland. Local.

Additional information
Flower appears to resemble an insect to the males of certain *hymenoptera*. The latter attempt to mate with the lower lip. In this way they assist in pollination. They do not, however, make clear distinctions between the different types of Ophrys and so hybrids frequently occur.

Orchid family *Orchidaceae*
Flowering time: June–July
Height of growth: 15–30 cm
Monocotyledonous; Perennial

Identification marks
1–10 flowers in a lax almost 1-sided spike. Marginal petals spread out, mauve to pink. Lip does not have spur, covered with velvety hairs and is longer than wide, convex, with a folded back 'beak' at the apex; brownish-red, with yellow markings. Deciduous leaves are narrow ovoid, alternate.

Habitat and distribution
Semi-dry turf, light areas in dry woodland. Very rare.

Additional information
Bee Orchids are very erratic in their appearance, producing abundant seed in some years then not reappearing for as long as a decade.

Man Orchid
Aceras anthropophorum

Lizard Orchid
Himantoglossum hircinum

Orchid family *Orchidaceae*
Flowering time: May–June
Height of growth: 20–30 cm
Monocotyledonous; Perennial

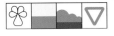

Identification marks
Inflorescence a narrow spike usually 5–15 cm long. Flowers do not have spur. Outer petals form a helmet shape approx. 5 mm across. They are green with a red or violet vein and margin. Lip approx. 1 cm long, divided into narrow lobes like the human body.

Habitat and distribution
Limy grassland, scrub and woodland. Thinly scattered though locally abundant in southern England.

Additional information
The scientific name (*anthropophorum* = carrying a man) indicates the shape.

Orchid family *Orchidaceae*
Flowering time: April–June
Height of growth: 20–80 cm
Monocotyledonous; Perennial

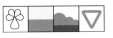

Identification marks
Inflorescence a very lax spike 15–25 cm long. Outer petals light green, forming a helmet shape, often joined together along edges, with conspicuous red veins and margins. Lip is trilobed: lateral lobes 5–7 mm long, usually brown. Median lobe 5–7 cm long, bilobed at the apex, twisted.

Habitat and distribution
Dry grassland and light woodland. Likes warmth. Very rare.

Additional information
In the Balkans there is the larger *H calcaratum*, the lip is narrower and much more deeply divided.

Green Figwort
Scrophularia umbrosa

Common Figwort
Scrophularia nodosa

Figwort family *Scrophulariaceae*
Flowering time: June–Aug.
Height of growth: 20–120 cm
Dicotyledonous; Perennial

Identification marks
Numerous small flowers in relatively dense pyramidal or ovoid panicles. Flowers approx. 7 mm long. Stem erect or ascending, broadly winged (wings measure a good 1/3 of the stem width). Leaves rounded at the base, small pinnate leaves absent from stem.
Habitat and distribution
Ditches.
Additional information
Similar: Water Figwort (*S. auriculata*), flowers barely 1 cm long. Stem erect, with winged angles (wings barely 1/6–1/5 of stem width, only conspicuous membraneous rib). Leaves lobed: lobes resemble pinnate leaflets. Common by ponds, streams and in damp woods.

Figwort family *Scrophulariaceae*
Flowering time: June–Sept.
Height of growth: 60–140 cm
Dicotyledonous; Perennial

Identification marks
Flowers in terminal panicle, approx. 8 mm long, dull brownish-red. Stem quadrangular but not winged. Leaves in opposite, alternate pairs and undivided.
Habitat and distribution
Woodland on ground rich in nutrients. Scattered.
Additional information
Very rarely there are mutants in the flower colour of the Figwort. The flowers are then yellow, more rarely white. These variations can easily be distinguished from the yellowish-green flowered Yellow Figwort. The Yellow Figwort (*S. vernalis*) flowers in spring and has its flowers in axillary panicles. Naturalized in a few localities.

Spindle
Euonymus europaeus

Buckthorn
Rhamnus catharticus

Spindle family *Celastraceae*
Flowering time: May
Height of growth: 1–3m
Dicotyledonous

Identification marks
Flowers usually have 4 petals, inconspicuous, green (without red spots) in small stalked axillary umbels. Branches green, 4 angled, leaves carried in opposite pairs are narrowly ovate and colour to a deep red in autumn. They are 3.5–5 cm in length. Capsule is carmine, 4-lobed, opening to show seeds with a fleshy orange-coloured aril.
Habitat and distribution
Scrub and hedgerows usually on chalk. Also on dunes. Widespread and frequent, except in northern Scotland.
Additional information
Many species of *Euonymus* are cultivated.

Buckthorn family *Rhamnaceae*
Flowering time: May–June
Height of growth: 4–6 m
Dicotyledonous

Identification marks
Branches and leaves opposite. Leaves petiolate, ovoid; margin finely toothed. Many small branches terminating in pointed thorns. Flowers in sparse axillary clusters on the previous year's growth, yellowish-green, inconspicuous, pleasant smelling. Black berries, pea-sized, inedible (or have purging effect).
Habitat and distribution
Hedges, dry thickets, oak and ash woods. Prefers calcareous soils. Scattered in England and Wales except for south-west areas, and Scotland.
Additional information
Similar: Rock Buckthorn, *R. saxatilis*, more delicate in every respect; leaves only 1–1.5 cm long. The Alder Buckthorn, *Frangula alnus* (see p. 379), has no thorns and leaves are alternate.

Alder Buckthorn
Frangula alnus

Bilberry
Vaccinium myrtillus

Buckthorn family *Rhamnaceae*
Flowering time: May–June
Height of growth: 1–4 m
Dicotyledonous

Identification marks
2–10 flowers in small axillary clusters approx.
4 mm across; greenish-white. Branches
always without thorns or thorny spikes.
Leaves ovate, margin entire, 4–7 cm long and
about 1/2 as wide, with 7–12 highly
prominent lateral veins on the undersides.
Berrylike drupe (only 1 seed), which is initially
red, then later black.
Habitat and distribution
Forest margins, clearings. Scattered.
Additional information
Like its close relative, Purging Buckthorn
(*Rhamnus catharticus*), the fruit has strongly
purgative properties.

Heath family *Ericaceae*
Flowering time: May–June
Height of growth: 15–40 cm
Dicotyledonous

Identification marks
Flowers solitary, occasionally in pairs in leaf
axils, rounded, opening shortly at the mouth,
greenish and usually tinged with red. Leaves
deciduous, elongate-ovate, pointed, slightly
crenate, green on both sides. Stem angular.
Habitat and distribution
Needs ground low in nutrients and virtually
free of lime. Common on heaths and moors
especially in the west and north.
Additional information
Similar: Bog Bilberry (*V. uliginosum*), flowers
rounded to ovoid, usually in terminal clusters.
Leaves blunt, with bluish-green undersides.
Stem round. Moors and peaty woodland.
Scattered. Berries edible in small quantities.

Mountain Currant
Ribes alpinum

Gooseberry
Ribes uva-crispa

Saxifrage family *Saxifragaceae*
Flowering time: April–June
Height of growth: 1–2.5 m
Dicotyledonous

Identification marks
20–30 male, 8–15 female flowers each in an erect raceme. Flowers usually dioecious, inconspicuous, greenish-yellow. Branches thornless. Leaves 2–4 cm long, 3–5 lobes. Lobes crenate. Berries red, insipid and slimy in taste.

Habitat and distribution
Needs stony calcareous ground. Grows in both dry and relatively moist woods on limestone, also on cliffs. Only in northern England and in Wales. Local.

Additional information
Mountain Currant is widespread, though never common in the mountains of Europe, as far east as Bulgaria and also in the Atlas in north Africa.

Saxifrage family *Saxifragaceae*
Flowering time: April–May
Height of growth: 50–120 cm
Dicotyledonous

Identification marks
1–3 axillary flowers; broadly campanulate, greenish-yellow, frequently tinged with red, with folded-back sepals. Branches have simple thorns, more rarely triple ones. Leaves round, 2–6 cm wide, sometimes in clusters 3–5 lobes. Fruit covered with relatively stiff hairs, green in the wild form, red or yellow in the cultivated form.

Habitat and distribution
Woods and hedges, frequent except in the far north of Scotland.

Additional information
There are numerous cultivated types. These originate in some cases from hybridization, hence growth forms and other characteristics may vary considerably. These may escape on to waste ground.

Sycamore
Acer pseudoplatanus

Ivy
Hedera helix

Maple family *Aceraceae*
Flowering time: April
Height of growth: 20–30 m
Dicotyledonous

Identification marks
Flowers pendulous in a raceme 5–15 cm long, containing up to 100 flowers. Flowers may be male, female or bisexual in one inflorescence. Flowers usually less than 1 cm across. Leaves 7–16 cm long, with 5 lobes. Lobes notched but not with a long point. Their stalk is usually reddish. The fruits are winged on one side.

Habitat and distribution
Found throughout Britain but never native. Self sows itself freely.

Additional information
Although introduced less than 400 years ago, the Sycamore behaves like a native tree. It grows best on deep moist soils where it makes a broad crowned tree, but also flourishes on moorland to 500 m.

Ivy family *Araliaceae*
Flowering time: Sept.–Nov.
Height of growth: 50 cm–30 m
Dicotyledonous

Identification marks
Flowers greenish-yellow, in large globular umbels in clusters at the ends of the climbing stems. Leaves evergreen. Lobed leaves on sterile growth along the ground or when climbing, but develop entire, ovate ones on flowering stems; the margin of these is entire. Fruit is a bluish-black berry.

Habitat and distribution
Likes half-shade and humid climate. Deciduous woodland, hedgerows. Scattered, locally in conspicuous abundance.

Additional information
The flowering and non-flowering branches of the Ivy plant differ considerably from one another.

Glossary

Achene Dry fruit which does not split open to release the single seed.

Amplexicaul With the base clasping the stem.

Apex Tip of a stem or other organ.

Ascending Directed upwards at a shallow angle.

Auricle Ear-shaped lobe or appendage at the base of a leaf.

Awn Stiff, bristle-like organ.

Axil The angle between the stem and a branch, leaf or bract.

Axillary Arising from an axil.

Balk Open grassy land.

Bifid Deeply notched or cleft.

Bipinnate A pinnate leaf with leaflets that are themselves pinnate.

Bract (Usually) modified leaf with a flower or inflorescence in its axil.

Bulb Underground storage organ formed of fleshy leaves or scales.

Calcareous Containing calcium carbonate.

Calcicolous Needing or preferring soils high in calcium carbonate.

Calcifugous Intolerant of soils high in calcium carbonate.

Calyx All the sepals of a flower (see Sepal).

Campanulate Shaped like a bell.

Capitulum Condensed head of flowers, especially of Compositae.

Capsule Dry fruit which splits open to release the seeds.

Cauline Borne on the stem.

Compound Leaf composed of a number of leaflets.

Cordate Heart-shaped.

Corolla All the petals of a flower (see Petal).

Corymb A raceme with the lower pedicels longer that the upper so that the flowers are all borne at more or less the same level.

Crenate With blunt or rounded teeth.

Cuneate Wedge-shaped.

Cyme An inflorescence formed by a stem terminating in a flower, subsequent flowers being produced below it so that the oldest flower is at the top (or centre) of the inflorescence.

Deciduous Shed annually, usually in the autumn.

Decurrent Extended downwards, as a leaf extending to form a wing on the stem.

Decussate Arranged in alternating pairs, each pair at 90° to the next.

Dentate Toothed.

Dichasium A cyme in which the first flower is flanked by 2 younger ones, which in turn are flanked by 2 even younger ones, and so on.

Dichotomous Regular; repeated division by forking into 2 equal branches.

Digitate Composed of 5 leaflets.

Dioecious With separate male and female flowers borne on different plants.

Distichous Arranged in 2 vertical rows.

Drupe Fleshy fruit with stony seed(s).

Entire Even, not toothed or lobed.

Epicalyx An additional whorl similar to, but outside, the true calyx.

Fascicle A bundle.

Fen carr Wet woodland growing over peat.

Follicle Dry fruit which splits open on 1 side only.

Fusiform Elongated, tapering at both ends.

Halophyte Plant able to tolerate salty conditions.

Hastate Spear-shaped.

Imparipinnate Leaf with the leaflets paired except for the terminal one.

Inflorescence The arrangement of all a plant's flowers.

Internode Section of stem between 2 nodes.

Involucre Whorl of bracts beneath an inflorescence or capitulum: adj. involucral.

Keel Boat-shaped lower petal(s) of a Pea flower.

Labiate Lipped, e.g. shaped like the flower of a Deadnettle.

Lamina The blade of a leaf.

Lanceolate Narrow and tapering at both ends.

Legume Seed-pod characteristic of the Pea family.

Ligule Strap-like extension of the corolla of a floret.

Linear Very narrow, with parallel sides.

-merous Having so many parts, e.g. 6-merous.

Monoecious With separate male and female flowers borne on the same plants.

Node Stem joint bearing a leaf.

Nut Dry fruit containing a single seed within a woody shell.

Obovate Inverted egg-shape, widest above the middle.

Ovary Female part of the flower containing the ovule(s) from which the seeds develop.

Ovate See Ovoid.

Ovoid Egg-shaped, widest below the middle.

Palmate Composed of more than 3 leaflets or lobes arising from a central point.

Panicle A branched raceme with sessile flowers.

Papillonaceous Butterfly-like, e.g. shaped like the flower of a Pea.

Pappus Parachute of hairs (sometimes reduced to scales) on a fruit, especially of Compositae.

Pedicel Individual flower stalk.

Peduncle Stalk of an inflorescence.

Peltate Circular; flat leaf with the petiole attached in the centre.

Perianth The sepals and petals together; especially when not distinguishable as calyx and corolla (see Tepal).

Petal One segment of a whorl of floral leaves, usually brightly coloured (see Corolla).

Petiolate Having a petiole (see below).

Petiole Stalk of a leaf.

Pinnate Leaf composed of several pairs of leaflets.

Prostrate Lying close to the ground.

Raceme An inflorescence formed by a stem producing a flower then growing further before producing the next, so that the youngest flowers are at the top of the stem; all pedicels are of equal length.

Radical Arising from the roots.

Receptacle Flat, domed or dished part of the stem to which the floral parts are attached.

Reniform Kidney-shaped.

Rhizome Underground, creeping stem producing roots and shoots.

Rhomboid Diamond-shaped.

Ruderal Plant typically found on wasteland and rubbish tips.

Sagittate Arrow-shaped.

Samara Dry fruit with part of the wall extended to form a wing.

Scape Leafless flower-stalk arising from the rootstock.

Scarious Dry, membranous.

Scorpioid Curved or coiled like the tail of a scorpion.

Scree Unstable slope of small, loose stones.

Sepal One segment of the whorl of floral parts (usually green) outside petals (see Calyx).

Serrate Sharply toothed like a saw.

Sessile Without a stalk.

Silicula Seed-pod which is broader than long, with a central partition which persists after the seeds are released; characteristic of some members of the cabbage family.

Silique Like a silicula but longer than broad; characteristic of the remaining members of the cabbage family.

Spadix Inflorescence in which sessile flowers are borne on a thickened, fleshy axis.

Spathe Large bract, often brightly coloured, which encloses a spadix.

Spatulate Spoon-shaped.

Spike A simple raceme with sessile flowers.

Stamen Male reproductive organ bearing pollen.

Standard The posterior petal of a Pea flower.

Stigma Part of the flower receptive to pollen.

Stipule Leaf-like organ at the base of a petiole.

Stolon Creeping overground stem rooting at intervals.

Style Usually elongated organ connecting the ovary and stigma.

Synanthropic Associated with man.

Tepal Term for 1 segment of the perianth when calyx and corolla are indistinguishable.

Trifoliate Composed of 3 leaflets.

Tuber Underground storage organ derived from a stem.

Umbel Umbrella-shaped inflorescence with all the pedicels arising from the same point.

Undulate Wavy.

Unilateral Arranged on 1 side.

Verticillaster Dichasial arrangement of flowers giving the appearance of a whorl.

Violaceous Flower shaped like that of a Pansy.

Wings The lateral petals of a Pea flower.

Index

Page numbers in *italics* indicate illustrations in addition to text.

Aaron's Rod *154*
Acer
  campestre *200*
  platanoides *200*
  pseudoplatanus *116, 381*
Aceras anthropophorum *261, 376*
Achillea
  millefolium *92, 244*
  ptarmica *92*
Acinos arvensis *345*
Aconite-leaved Buttercup *57*
Aconitum
  lycotonum *180*
  napellus *331*
  vulparia *180*
Acorus calamus *352*
Actaea spicata *35*
Adenostyles
  alliariae *242*
  glabra *242*
Adonis
  aestivalis *240*
  autumnalis *240*
  vernalis *158*
Adoxa moschatellina *369*
Aegopodium podagraria *67*
Agrimonia eupatoria *142*
  ssp. grandis *142*
Agrimony *142*
Ajuga
  genevensis *341*
  pyramidalis *341*
  reptans *341*
Alchemilla vulgaris *360*
Alder Buckthorn *199, 378, 379*
Alectorolophus minor *195*
Alisma
  gramineum *32, 206*
  lanceolatum *32, 206*
  plantago-aquatica *32, 206*
Alliaria
  officinalis *36*
  petiolata *36*
Allium
  sphaerocephalum *239*
  ursinum *80*
  victorialis *371*
  vineale *239*
Alpine Avens *139*

Alpine Bartsia *345*
Alpine Bellflower *316*
Alpine Butterbur *243*
Alpine Butterwort *103*
Alpine Cinquefoil *140*
Alpine Columbine *304*
Alpine Enchanter's-nightshade
  44
Alpine Fireweed *211*
Alpine Leek *371*
Alpine Pasqueflower *57*
Alpine Sainfoin *270*
Alpine Snowbell *326*
Alpine Sow-thistle *329*
Alpine Squill *323*
Alpine Thistle *245*
Alpine Toadflax *346*
Alsike Clover *99, 264*
Alternate-leaved Golden
  Saxifrage *125*
Alyssum
  alyssoides *122*
  calycinum *122*
  montanum *122*
  saxatile *122*
Amaranthus
  albus *359*
  hybridus *359*
  retroflexus *359*
Amelanchier ovalis *107*
Anacamptis pyramidalis *253*
Anagallis
  arvensis *231, 306*
    ssp. foemina *306*
  foemina *306*
Anchusa arvensis *314*
Andromeda polifolia *288*
Androsace lactea *72*
Anemone
  dubia *56*
  narcissiflora *56*
  nemorosa *85*
  ranunculoides *133*
  sylvestris *57*
Angelica
  archangelica *368*
  sylvestris *69, 229*
Angular Solomon's-seal *81*
Annual Knawel *366*
Annual Mercury *363*
Annual Woundwort *101*
Antennaria
  carpatica *244*
  dioica *244*

Anthemis arvensis *90*
Anthericum
  lilago *79*
  ossifragum *156*
  ramosum *79*
Anthriscus sylvestris *68*
Anthyllis vulneraria *184*
  ssp. alpestris *184*
  ssp. maritima *184*
  ssp. vulneraria *184*
Aphanes
  arvensis *360*
  microcarpa *36*
Apple *286*
Aquilegia
  alpina *304*
  atrata *304*
  einseleana *304*
  vulgaris *304*
Arabidopsis thaliana *39*
Arabis glabra *38, 121*
Arctium
  lappa *250*
  minus *250*
  nemorosum *250*
  tomentosum *249*
Arctostaphylos uva-ursi *113, 114,*
  289
Aristolochia clematis *180*
Armeria maritima *232*
Arnica *165*
Arnica montana *165*
Arrowhead *32*
Artemisia vulgaris *164*
Arum maculatum *34, 352*
Aruncus
  dioicus *64*
  sylvestris *64*
  vulgaris *64*
Asarabacca *353*
Asarum europaeum *353*
Asperula odorata *45*
Aster
  amellus *326*
  tripolium *326*
Astragalus glycyphyllos *182*
Atriplex
  glabriuscula *358*
  hastata *358*
  laciniata *358*
  littoralis *358*
  patula *358*
Atropa bella-donna *236,*
  369

Autumn Hawkbit 171
Autumn Squill 323
Avens
  Alpine 139
  Creeping 139
  Mountain 86
  Water 222, 368
  Wood 139

Babington's Orache 358
Ball Mustard 123
Ballota nigra 274
  ssp. foetida 274
  ssp. nigra 274
Balsam
  Indian 271
  Small 189
  Touch-me-not 189
Baneberry 35
Barbarea
  intermedia 120
  stricta 120
  vulgaris 120
Barberry 202
Barren Strawberry 63
Bartsia alpina 345
Basil Thyme 345
Bastard Balm 102, 272
Bearberry 113, 114, 289
Bearded Bellflower 316
Bear's-ear 148
Bedstraw
  Common Marsh 47
  Fen 47
  Heath 47
  Hedge 47
  Our Lady's 45
  Round-leaved 46
  Wood 46
Bee Orchid 375
Bellflower
  Alpine 316
  Bearded 316
  Clustered 319
  Creeping 318
  Large 318
  Nettle-leaved 318
  Peach-leaved 319
  Spreading 317
  Yellow 155
Bellis perennis 91
Berberis vulgaris 202
Betonica
  hirsuta 276

officinalis 276
stricta 276
Betony 276
Bidens
  radiata 163
  tripartita 163
  trondosa 163
Bilberry 289, 379
Bindweed
  Black 48
  Copse 48
  Hairy 74
  Hedge 74
  Large 74
  Lesser 233
  Sea 74
Bird Cherry 111
Bird's-eye Primrose 230
Bird's-nest Orchid 179, 373
Birthwort 180
Biscutella laevigata 124
Bishop's Weed 67
Biting Stonecrop 138
Bitter Milkwort 336
Bitter-cress
  Drooping 119
  Hairy 40
  Large 41
  Narrow-leaved 40, 208, 294
  Wavy 40
Bittersweet 315
Black Bindweed 48
Black Broom 204
Black Horehound 274
Black Medick 183
Black Mustard 121
Black Nightshade 75
Black Pea 268
Blackthorn 111
Bladder Campion 55
Bloody Crane's-bill 226
Blue Lettuce 329
Blue Pimpernel 306
Blue-spiked Rampion 320
Blue Water Speedwell 297
Bog Arum 34
Bog Asphodel 156
Bog Bilberry 113, 289, 379
Bog Rosemary 288
Bog Violet 337
Bogbean 73
Box 198
Bramble 61, 112
Branched Bur-reed 118, 350

Branched St Bernard's Lily 79
Brassica nigra 121
Breckland Thyme 278
Broad-leaved Dock 355
Broad-leaved Helleborine 94, 372
Broad-leaved Pondweed 351
Broad-leaved Spurge 128
Broad-leaved Willowherb 213
Brooklime 297
Broom 204
Brown Gentian 152
Brown Knapweed 248
Bryonia cretica ssp. dioica 77
Buck's-horn Plantain 45
Buckthorn 199, 378
  Alder 199, 378, 379
  Purging 379
  Rock 199, 378
Bugle 341
Bugloss 314
Buglossoides
  arvensis 74
  purpurocaeruleum 313
Bulbous Buttercup 135
Buphthalmum salicifolium 162
Bupleurum falcatum 147
Bur-reed
  Branched 118, 350
  Floating 118, 350
  Least 118, 350
  Unbranched 118, 350
Burdock
  Downy 249
  Greater 250
  Lesser 250
Burnet-saxifrage 69
Burning Bush 271
Burnt Orchid 96, 255
Bush Vetch 334
Butomus umbellatus 238
Butterbur 88, 243
  Alpine 243
  White 88
Buttercup
  Aconite-leaved 57
  Bulbous 135
  Corn 134
  Creeping 135
  Goldilocks 133
  Hairy 135
  Large White 57
  Meadow 134
  Wood 134, 136
  Woolly 136

Buxux sempervirens 198

Cabbage Thistle 170
Cakile maritima 37, 209
Calamintha acinos 345
Calendula
 arvensis 169
 officinalis 169
Calla palustris 34
Callitriche palustris 364
Calluna vulgaris 291
Caltha palustris 132
Calystegia
 pulchra 74
 sepium 74
 silvatica 74
 soldanella 74
Cambridge Milk-parsley 69
Camelina sativa 123
Camomilla recutita 90
Campanula
 alpina 316
 barbata 316
 cervicaria 319
 glomerata 319
 latifolia 318
 patula 317
 persicifolia 319
 rapunculoides 318
 rotundifolia 317
 thyrsoides 155
 trachelium 318
Campion
 Bladder 55
 Moss 219
 Red 219
 White 54
Canadian Fleabane 88
Canadian Goldenrod 160
Capsella bursa-pastoris 42
Caraway 67
Cardamine
 amara 41
 bulbifera 209, 294
 enneaphyllos 119
 flexuosa 40
 hirsuta 40
 impatiens 40, 208, 294
 pratensis 40, 208, 294
Cardaria draba 39
Carduus
 defloratus 245
 nutans 245
 platylepis 245

Carlina
 acaulis 91
 vulgaris 169
Carline Thistle 169
Carpathian Cat's-foot 244
Carthusian Pink 216
Carum carvi 67
Catchfly
 Forked 54
 Night-flowering 54
 Nottingham 54
 Sticky 220
Cat's-ear 171
Celery-leaved Crowfoot 136
Centaurea
 cyanus 327
 jacea 248
 montana 327
 nigra 248, 327
  ssp. nemoralis 248
  ssp. nigra 248
 scabiosa 249, 327
Centaurium
 erythraea 233
 pulchellum 233
Cephalanthera
 alba 95
 damasonium 95
 grandiflora 95
 latifolia 95
 longifolia 95
 pallens 95
 rubra 252
Cerastium
 arvense 52
 brachypetalum 52
 caespitosum 52
 fontanum ssp. triviale 52
 glomeratum 52
 holosteoides 52
 semidecandrum 52
 tomentosum 52
 triviale 52
Cerasus avium 110
Cerinthe glabra 152
Chaenorhinum minus 347
Chaerophyllum aureum 68
Chamaenerion 211
Chamaespartium sagittale 181
Chamomilla suaveolens 163
Charlock 38, 120
Cheddar Pink 217
Cheiranthus cheiri 121
Chelidonium majus 119

Chenopodium
 album 357
 bonus-henricus 357
Cherry
 Bird 111
 Cornelian 199
 Garden 110
 Wild 110
Chickweed 50, 53, 65
Chickweed Wintergreen 87
Chicory 328
Christmas Rose 56
Chrysanthemum
 corymbosum 93
 leucanthemum 93
 tanacetum 164
 vulgare 164
Chrysosplenium
 alternifolium 125
 oppositifolium 125
Cicerbita alpina 329
Cichorium intybus 328
 ssp. foliosum 328
 ssp. sativa 328
Cinquefoil
 Alpine 140
 Creeping 141
 Golden 140
 Marsh 223
 White 63
Circaea
 alpina 44
 intermedia 44
 lutetiana 44
Cirsium
 acaule 247
 arvense 328
 caulescens 247
 hypoleucum 246
 oleraceum 170
 palustre 246
 spinosissimum 170
 vulgare 246
Cleavers 47
Clematis vitalba 104
Clover
 Alsike 99, 264
 Crimson 265
 Hare's-foot 98
 Mountain 99
 Persian 265
 Red 264
 White 98
 Zigzag 264

Clustered Bellflower 319
Clustered Dock 354
Coeloglossum viride 373
Colchicum
    alpinum 238
    autumnale 238
    bulbocodium 238
Colt's-foot 88, 165
Columbine 304
Columbine-leaved Meadow
    Rue 292
Comarum palustre 223
Common Amaranthus 359
Common Bird's-foot-trefoil 186
Common Bistort 215
Common Butterwort 103, 347
Common Centaury 233
Common Comfrey 153, 234, 310
Common Cornsalad 315
Common Cow-wheat 193
Common Cyclamen 232
Common Dodder 234
Common Dog Violet 338
Common Duckweed 351
Common Evening-primrose
    130
Common Field Speedwell
    299
Common Figwort 377
Common Frogbit 33
Common Fumitory 263
Common Gromwell 74
Common Hempnettle 273
Common Honesty 293
Common Knapweed 248
Common Mallow 228
Common Marsh Bedstraw 47
Common Meadow-rue 118
Common Milkwort 270, 336
Common Mouse-ear 52
Common Nettle 354
Common Orache 358
Common Poppy 208
Common Ragwort 168
Common Rock-rose 143
Common Sainfoin 266
Common Sorrel 207
Common Stork's-bill 223
Common Toadflax 197
Common Twayblade 372
Common Valerian 237
Common Vetch 266
Common Water-crowfoot 58
Common Whitebeam 108

Common Whitlow-grass 39
Common Wintergreen 71
Conringia orientalis 38
Consolida
    ajacis 330
    regalis 330
Convallaria najalis 82
Convolvulus arvensis
    233
Conyza canadensis 88
Copse Bindweed 48
Coral-root Orchid 97
Corallorhiza
    innata 97
    trifida 97
Coralroot 209, 294
Corn Buttercup 134
Corn Chamomile 90
Corn Cleavers 47
Corn Mint 296
Corn Spurrey 53
Cornelian Cherry 199
Cornflower 327
Cornus
    mas 199
    sanguinea 105
Coronilla varia 263
Corydalis
    cava 97, 262
    intermedia 262
    lutea 181
    solida 97, 262
Cotoneaster
    integerrimus 286
    tomentosus 286
Cow Parsley 68
Cow-wheat
    Common 193
    Field 279
    Small 193
Cowberry 113, 289
Cowslip 149
Crab Apple 106, 286
Crambe maritima 37
Cranberry 283
Crane's-bill
    Bloody 226
    Cut-leaved 224
    Dusky 227
    Hedgerow 225
    Long-stalked 224
    Marsh 226
    Meadow 305
    Wood 227, 305

Crataegus
    laevigata 109
    monogyna 109
Creeping Avens 139
Creeping Bellflower 318
Creeping Buttercup 135
Creeping Cinquefoil 141
Creeping Gypsophila 215
Creeping Jenny 151
Creeping Lady's-tresses
    95
Creeping Restharrow 291
Creeping Spearwort 137
Creeping Thistle 328
Crepis
    biennis 175
    capillaris 175
    mollis 175
    virens 175
Cress
    Hoary 39
    Shepherd's 39
    Thale 39
Crimson Clover 265
Crocus vernus 84, 324
    ssp. albiflorus 84
    ssp. vernus 324
Cross Gentian 295
Cross-leaved Heath 284
Crosswort 131
Crow Garlic 239
Crowberry 288
Crown Vetch 263
Cruciata
    chersonensis 131
    laevipes 131
Cuckoo Flower 40, 208, 294
Cuckoo-pint 34, 352
Curled Dock 355
Cuscuta
    epithymum 234
    europaea 234
    pilinum 234
Cut-leaved Crane's-bill 224
Cut-leaved Deadnettle 275
Cyclamen purpurascens 232
Cymbalaria muralis 346
Cynanchum vincetoxicum 73
Cypress Spurge 129, 362
Cypripedium calceolus 178
Cytisus
    nigricans 204
    sagitalis 181
    scoparius 204

Dactylorhiza
  incarnata 259
  maculata 258
  majalis 259
  sambucina 179, 258
Daisy 91
Dame's Violet 293
Dandelion 172
Danewort 76
Daphne
  cneorum 283
  mezereum 282
  ptraea 282
  striata 282
Dark Mullein 154
Dark Rampion 320
Dark Red Helleborine 253
Daucus carota 69
Deadly Nightshade 236, 369
Deadnettle
  Cut-leaved 275
  Red 275
  Spotted 274
  White 100⁻
Dentaria
  bulbifera 294
  enneaphyllos 119
Deptford Pink 216
Devil's-bit Scabious 303
Dewberry 112
Dianthus
  armeria 216
  barbatus 216
  carthusianorum 216
  deltoides 217
  gratianopolitanus 217
  plumarius 218, 304
  seguieri 217, 218
  superbus 218, 304
Dictamnus albus 271
Digitalis
  grandiflora 196
  lutea 196
  x *purpurascens* 196
  purpurea 280
Dipsacus
  fullonum 303
  laciniatus 33
Dock
  Broad-leaved 355
  Clustered 354
  Curled 355
  Golden 355

Great Water 355
  Wood 354
Dodder 234
Dog Rose 287
Dog's Mercury 363
Dogwood 105
Dotted Loosestrife 150
Downy Burdock 249
Downy Hemp-nettle 191
Draba aizoides 124
Dragon's-tooth 187
Drooping Bitter-cress 119
Drooping Star-of-Bethlehem 78
Dropwort 64, 87
Drosera
  anglica 59
  intermedia 59
  obovata 59
  rotundifolia 59
Dryas octopetala 86
Duckweed
  Common 351
  Gibbous 351
  Ivy 351
Dusky Crane's-bill 227
Dwarf Elder 76
Dwarf Mallow 229
Dwarf Milkwort 336
Dwarf Snowbell 241
Dwarf Spurge 129
Dyer's Greenweed 203

Early Dog Violet 338
Early Goldenrod 160
Early Marsh Orchid 259
Early Purple Orchid 257
Early Spider Orchid 374
Echium
  italicum 340
  vulgare 340
Elder 76, 115
  Dwarf 76
  Ground 67
  Red-berried 201
Elder-flowered Orchid 179, 258
Elm
  English 285
  Small-leaved 285
  Wych 285
Empetrum nigrum 288
Enchanter's-nightshade 44
English Elm 285

Epilobium
  alpestre 210
  angustifolium 211
  dodonaei 211
  fleischeri 211
  hirsutum 212
  montanum 213
  parviflorum 212
  roseum 210
Epipactis
  atrorubens 253
  helleborine 94, 372
  palustris 94
  purpurata 253, 372
Epipogium aphyllum 96
Erica
  carnea 284
  herbacea 284
  tetralix 284
Erigeron canadensis 88
Erodium cicutarium 223
Erophila verna 39
Eruca sativa 121
Eryngium
  campestre 66
  maritimum 306
Erysimum cheiranthoides 121
Erythronium dens-canis 232
Euonymus europaeus 378
Eupatorium cannabinum 242
Euphorbia
  amygdaloides 127
  brittingeri 127
  cyparissias 129, 362
  dulcis 362
  esula 129
  exigua 129
  falcata 129
  helioscopia 128, 362
  peplis 128, 362
  platyphyllos 128
  verrucosa 127
Euphrasia rostkoviana 103
Eyebright 103

Fallopia
  convolvulus 48
  dumetorium 48
False Acacia 117
False Helleborine 78, 152
Fan-leaved Water-crowfoot 58
Fat Hen 357
Fen Bedstraw 47
Fen Violet 337

Ficaria verna 158
Field Cow Wheat 279
Field Eryngo 66
Field Forget-me-not 312
Field Gromwell 74
Field Madder 302
Field Maple 200
Field Marigold 169
Field Mouse-ear 52
Field Penny-cress 42
Field Rose 110
Field Scabious 349
Figwort
  Common 377
  Green 377
  Water 377
  Yellow 377
Filipendula
  hexapetala 87
  ulmaria 64
  vulgaris 64, 87
Fine-leaved Vetch 335
Fingered Speedwell 299
Fireweed 211
Fleabane 161
Floating Bur-reed 118, 350
Flowering Rush 238
Fly Honeysuckle 116, 205
Fly Orchid 375
Forget-me-not
  Field 312
  Water 312
  Wood 312, 313
Forked Catchfly 54
Forking Larkspur 330
Foxglove 280
  Large Yellow 196
  Small Yellow 196
Fragaria
  moschata 62
  vesca 62
Fragrant Orchid 260
Franġula alnus 199, 378, 379
Fringed Gentian 295
Fringed Water-lily 33
Fritillaria meleagris 82, 370
Fritillary 82, 370
Frog Orchid 373
Frosted Orache 358
Fumaria
  officinalis 263
  parviflora 263
  vaillantii 263
Furze 22

Gagea
  lutea 157
  pratensis 157
  villosa 157
Galanthus nivalis 83
Galeobdolon luteum 192
Galeopsis
  pubescens 273
  segetum 191
  speciosa 191, 273
  tetrahit 273
Galinsoga
  ciliata 89
  parviflora 89
Galium
  aparine 47
  cruciata 131
  mollugo 47
  odoratum 45
  palustre 47
  x pomeranicum 131
  rotundifolium 46
  saxatile 47
  scabrum 46
  sylvaticum 46
  tricornutum 47
  uliginosum 47
  verum 131
Gallant Soldier 89
Garden Angelica 368
Garden Cherry 110
Garden Privet 105
Garlic Mustard 36
Genista
  anglica 203
  germanica 203
  pilosa 203
  sagitalis 181
  tinctoria 203
Gentian
  Brown 152
  Cross 295
  Fringed 295
  Great Yellow 78, 152
  Marsh 308
  Milkweed 308
  Purple 152
  Spotted 152
  Spring 309
  Stemless 309
Gentiana
  acaulis 309
  asclepiadea 308
  clusii 309

  cruciata 295
  lutea 78, 152
  pannonica 152
  pneumonanthe 308
  punctata 152
  purpurea 152
  verna 309
Gentianella cilata 295
Geranium
  columbinum 224
  dissectum 224
  palustre 226
  phaeum 227
  pratense 305
  purpureum 225
  pyrenaicum 225
  robertianum 225
    ssp. maritimum 225
  sanguineum 226
  sylvaticum 227
  sylvaticus 305
German Asphodel 156
German Greenweed 203
Germander Speedwell 301
Geum
  montanum 139
  reptans 139
  rivale 222, 368
  urbanum 139
Ghost Orchid 96
Giant Hogweed 368
Gibbous Duckweed 351
Gladiolus
  imbricatus 252
  palustris 252
Glasswort 356
Glaux maritima 231
Glechoma hederacea 340
Globe-flower 159
Globularia
  nudicaulis 348
  punctata 348
Gnaphalium silvaticum 89
Goat's beard 172
Goat's-beard Spiraea 64
Gold-of-pleasure 123
Golden Alison 122
Golden Chervil 68
Golden Cinquefoil 140
Golden Dock 355
Goldenrod 160
Goldilocks Buttercup 133
Good King Henry 357
Goodyera repens 95

Gooseberry 380
Gorse 202
Goutweed 67
Grass-leaved Orache 358
Grass-of-Parnassus 60
Gratiola officinalis 102
Great Burnet 210, 361
Great Butterfly-orchid 94
Great Plantain 365
Great Spearwort 137
Great Sundew 59
Great Water Dock 355
Great Willowherb 212
Great Yellow Gentian 78, 152
Greater Bird's-foot'trefoil 186
Greater Bladderwort 197
Greater Bur-marigold 163
Greater Burdock 250
Greater Burnet-saxifrage 69
Greater Celandine 119
Greater Knapweed 249
Greater Periwinkle 307
Greater Reedmace 350
Greater Stitchwort 51
Greater Yellow Rattle 195
Green Figwort 377
Green Hellebore 367
Green-winged Orchid 254
Greenweed
    Dyer's 203
    German 203
    Hairy 203
Grey Field Speedwell 298
Grey Mouse-ear 52
Gromwell
    Common 74
    Field 74
    Purple 313
Ground Elder 67
Ground Ivy 340
Groundsel 166
Guelder-rose 116
Gymnadenia conopsea 260
Gypsophila
    muralis 215
    repens 215

Hairy Alpenrose 290
Hairy Bindweed 74
Hairy Bitter-Cress 40
Hairy Buttercup 135
Hairy Greenweed 203
Hairy St John's-wort 144
Hairy Tare 334

Hairy Violet 339
Harebell 317
Hare's-ear Mustard 38
Hare's-foot Clover 98
Hautbois Strawberry 62
Hawk's-beard
    Northern 175
    Rough 175
    Smooth 175
Hawkweed
    Mouse-ear 176
    Orange 251
    Umbellate group 177
Hawkweed Oxtongue 171
Hawthorn 109
Heart's-ease 190
Heath Bedstraw 47
Heath Cudweed 89
Heath Dog Violet 338
Heath Groundsel 166
Heath Milkwort 337
Heath Speedwell 300
Heath Spotted Orchid 258
Heather 291
Hedera helix 381
Hedge Bedstraw 47
Hedge Bindweed 74
Hedge Garlic 36
Hedge Hyssop 102
Hedge Mustard 121
Hedge Woundwort 277
Hedgerow Crane's-bill 225
Hedysarum hedysaroides 270
Helianthemum nummularium
    143
    ssp. glabrum 143
    ssp. grandiflorum 143
    ssp. nummularius 143
Helianthus tuberosus 162
Helleborine
    Broad-leaved 94, 372
    Dark Red 253
    False 78, 152
    Marsh 94
    Narrow-leaved 95
    Red 252
    White 95
Helleborus
    foetidus 367
    niger 56
    viridus 367
Hemp Agrimony 242
Henbane 153
Henbit 275

Hepatica 325
Hepatica nobilis 325
Heracleum
    mantegazzianum 368
    sphondylium 68
Herb Gerard 67
Herb Paris 371
Herb Robert 225
Herminium monorchis 373
Hesperis matronalis 293
Hieracium
    aurantiacum 251
    caespitosum 251
    murorum group 177
    pilosella 176
    sabauda group 177
    sylvaticum 177
    umbellatum 177
Himantoglossum
    calcaratum 261, 376
    hircinum 261, 376
Hippocrepis comosa 187
Hippuris vulgaris 364
Hoary Cress 39
Hoary Plaintain 302, 365
Hoary Ragwort 168
Hoary Willowherb 212
Hogweed 68
Holly 104
Homogyne
    alpina 243
    discolor 243
Honeysuckle 117
Hop 353
Hop Trefoil 185
Horseshoe Vetch 187
Hottonia palustris 72
Houseleek 240
Humulus lupulus 353
Hydrocharis morsus-ranae
    33
Hydrocotyle vulgaris 70
Hyoscyamus niger 153
Hypericum
    acutum 145
    x desetangsii 145
    hirsutum 144
    humifusum 143
    maculatum 145
    montanum 144
    perforatum 146
    pulchrum 146
    tetrapterum 145
Hypochoeris radicata 171

Ilex aquifolium 104
Impatiens
  balfourii 271
  glandulifera 271
  noli-tangere 189
  parviflora 189
Imperforate St John's-wort 145
Indian Balsam 271
Intermediate Winter-green 71
Inula
  conyza 161
  salicina 161
Iris
  pseudacorus 157
  sibirica 323
Irish Fleabane 161
Irish Marsh Orchid 259
Isatis tinctoria 123
Ivy 381
Ivy Duckweed 351
Ivy-leaved Speedwell 298
Ivy-leaved Toadflax 346

Jack-by-the-Hedge 36
Jack-in-the-pulpit 34, 352
Jasione montana 321
Jerusalem Artichoke 162
Jupiter's Distaff 192

Kidney Vetch 184
Knapweed
  Brown 248
  Common 248
  Greater 249
  Mountain 327
Knautia
  arvensis 349
  dipsacifolia 349
Knotgrass 49, 214

Labrador-tea 114
Lactuca
  muralis 174
  perennis 329
  scariola 176
  serriola 176
Lady Orchid 256
Lady's Mantle 360
Lady's-slipper 178
Lamiastrum galeobdolon 192
  ssp. flavidum 192
  ssp. galeobdolon 192
  ssp. montanum 192
Lamium

album 100
amplexicaule 275
galeobdolon 192
hybridum 275
maculatum 274
purpureum 275
Lapsana communis 174
Large Bellflower 318
Large Bindweed 74
Large Bitter-cress 41
Large-flowered Hemp-nettle 191,
  273
Large-flowered Mullein 155
Large Hop Trefoil 184
Large Self-heal 343
Large White Buttercup 57
Large Yellow Foxglove 196
Lathraea squamaria 281
Lathyrus
  aphaca 188
  japonicus 267
  liniifolius 269, 333
  niger 268
  pratensis 188
  sylvestris 267
    ssp. angustifolius 267
    ssp. platyphyllos 267
    ssp. sylvestris 267
  tuberosus 268
  vernus 269, 333
Leafy Lousewort 194
Leafy Spurge 129
Least Bur-reed 118, 350
Least Water-lily 132
Ledum
  groenlandicum 114
  palustre 114
Leek
  Alpine 371
  House 240
  Round-headed 239
Legousia
  hybrida 3200
  speculum-veneris 320
Lembotropis nigricans 204
Lemna
  gibba 351
  minor 351
  trisulca 351
Leontodon
  autumnalis 171
  hispidus 171
Lesser Bindweed 233
Lesser Burdock 250

Lesser Butterfly-orchid 94
Lesser Celandine 158
Lesser Grape Hyacinth 322
Lesser Periwinkle 307
Lesser Reedmace 350
Lesser Skullcap 342
Lesser Spearwort 137
Lesser Stitchwort 50
Lesser Trefoil 185
Lesser Twayblade 372
Lettuce
  Blue 329
  Prickly 176
  Red Hare's 251
  Wall 174
Leucanthemum
  corymbosum 93
  vulgare 93
Leucojum
  aestivum 83
  vernum 83
Ligustrum
  ovalifolium 105
  vulgare 105
Lilium martagon 239
Lily
  Branched St Bernard's 79
  Martagon 239
  May 35
  St Bernard's 79
Lily-of-the-valley 82
Limestone Woundwort 276
Limodore 330
Limodorum abortivum 330
Limonium vulgare 307
Linaria
  alpina 346
  minor 347
  vulgaris 197
Linum catharticum 65
Listera
  cordata 372
  ovata 372
Lithospermum
  arvense 74
  officinale 74
Little Mouse-ear 52
Livelong Saxifrage 61
Lizard Orchid 261, 376
Long-flowered Primrose 230
Long-headed Poppy 208
Long-stalked Crane's-bill 224
Lonicera
  caprifolium 117

periclymenum 117
ruprechtiana 116, 2005
xylosteum 116, 205
Loosestrife
Dotted 150
Purple 241
Slender 241
Tufted 150
Yellow 150
Lords and Ladies 34, 352
Lotus
corniculatus 186
pedunculatus 186
siliquosus 187
uliginosus 186
Lousewort 281
Leafy 194
Marsh 281
Whorled 280
Lucerne 332
Lunaria
annua 293
rediviva 293
Lungwort 235, 311
Lupinus polyphyllus 331
Lychnis
flos-cuculi 220
viscaria 220
Lycopsis arvensis 314
Lysimachia
nemorum 151
nummularia 151
punctata 150
thyrisiflora 150
vulgaris 150
Lythrum
salicaria 241
virgatum 241

Mahonia aquifolium 104
Maianthemum bifolium 35
Maiden Pink 217
Malus
domestica 106, 286
sylvestris 106, 286
Malva
alcea 228
moschata 228
neglecta 229
pusilla 229
sylvestris 228
Man Orchid 261, 376
Mare's-tail 364
Marguerite 93

Marjoram 278
Marsh Cinquefoil 223
Marsh Crane's-bill 226
Marsh Gentian 308
Marsh Gladiolus 252
Marsh Helleborine 94
Marsh Lousewort 281
Marsh-marigold 132
Marsh Pennywort 70
Marsh Stitchwort 51
Marsh Thistle 173, 246
Marsh Valerian 237
Marsh Woundwort 277
Martagon Lily 239
Matricaria
corymbosum 93
inodora 90
maritimum 90
matricarioides 163
perforata 90
suaveolens 163
May Lily 35
Meadow Buttercup 134
Meadow Crane's-bill 305
Meadow Saffron 238
Meadow Sage 344
Meadow Saxifrage 60
Meadow Vetchling 188
Meadowsweet 64
Medicago
falcata 332
lupulina 183
sativa 332
ssp. falcata 183
Medium-flowered Winter-cress
120
Medlar 107
Melampyrum
arvense 279
pratense 193
sylvaticum 193
Melandrium
album 54
dioicum 219
rubrum 219
Melilotus
alba 99
altissima 182
officinalis 182
Melittis melissophyllum 102,
272
Mentha
aquatica 213, 296
arvensis 296

x piperita 213
spicata 296
Menyanthes trifoliata 73
Mercurialis
annua 363
perennis 363
Merspilus germanica 107
Mezereon 282
Midland Hawthorn 109
Military Orchid 255
Milkweed Gentian 308
Milkwhite Rock-jasmine 72
Milkwort
Bitter 336
Common 270, 336
Dwarf 336
Heath 337
Sea 231
Shrubby 205
Tufted 270
Mistletoe 198
Moehringia trinervia 53
Moneses uniflora 70
Monkey Orchid 257
Monk's-hood 331
Monk's Rhubarb 356
Monotropa hypopitys 130
ssp. hypophegea 130
ssp. hypopitys 130
Moon Daisy 93
Moor-king 194
Moschatel 369
Moss Campion 219
Moth Mullein 154
Mountain Ash 109
Mountain Aster 326
Mountain Avens 86
Mountain Clover 99
Mountain Currant 380
Mountain Everlasting 244
Mountain Knapweed 327
Mountain Valerian 236
Mouse-ear
Common 52
Field 52
Grey 52
Little 52
Sticky 52
Mouse-ear Hawkweed
176
Mugwort 164
Mullein
Dark 154
Large-flowered 155

Moth 154
White 75, 154
Muscari
  botryoides 322
  comosum 322
  neglectum 322
  racemosum 322
Musk Mallow 228
Musk Orchid 373
Musk Thistle 245
Mustard
  Ball 123
  Black 121
  Garlic 36
  Hare's-ear 38
  Hedge 121
  Tower 38, 121
  Treacle 121
Mycelis muralis 174
Myosotis
  alpestris 313
  arvensis 312
  caespitosa 312
  scorpioides 312
  secunda 312
  sylvatica 312, 313
Myosoton aquaticum 51

Narcissus-flowered Anemone 56
Narrow-leaved Bitter-cress 40,
  208, 294
Narrow-leaved Everlasting Pea
  267
Narrow-leaved Helleborine 95
Narrow-leaved Water Plantain
  32, 206
Narrow-leaved Yellow-rattle 195
Narthecium ossifragum 156
Nasturtium
  microphyllum 41
  officinale 41
Neottia nidus-avis 179, 373
Neslia paniculata 123
Nettle-leaved Bellflower 318
Night-flowering Catchfly 54
Nightshade
  Alpine Enchanter's- 44
  Black 75
  Deadly 236, 369
  Enchanter's- 44
  Upland Enchanter's- 44
Nigritella nigra 260
Nipplewort 174
Northern Hawk's-beard 175

Norway Maple 200
Nottingham Catchfly 54
Nuphar
  lutea 132
  luteum 84
  pumila 132
Nymphaea alba 84
Nymphoides peltata 33

Oblong-leaved Sundew 59
Odontites verna
  ssp. *serotina* 279
  ssp. *verna* 279
Oenothera
  biennis 130
  parviflora 130
One-flowered Wintergreen 70
One-rowed Watercress 41
Onobrychis viciifolia 266
Ononis
  repens 291
  spinosa 291
Ophrys
  apifera 375
  holosericea 374
  insectifera 375
  sphegodes 374
Opposite-leaved Golden
Saxifrage 125
Orange Hawkweed 251
Orchid
  Bee 375
  Bird's-nest 179, 373
  Burnt 96, 255
  Coral-root 97
  Early Marsh 259
  Early Purple 257
  Early Spider 374
  Elder-flowered 179, 258
  Fly 375
  Fragrant 260
  Frog 373
  Ghost 96
  Great Butterfly- 94
  Green-winged 254
  Heath Spotted 258
  Irish Marsh 259
  Lady 256
  Lesser Butterfly- 94
  Lizard 261, 376
  Man 261, 376
  Military 255
  Monkey 257
  Musk 373

Pale-flowered 178
Pyramidal 253
Round-headed 254
Vanilla 260
Orchid family 374
Orchis
  laxiflora
    ssp. *laxiflora* 256
    ssp. *palustris* 256
  mascula 257
  militaris 255
  morio 254
  pallens 178
  purpurea 256
  simia 257
  ustulata 96, 255
Oregon Grape 104
Origanum vulgare 278
Ornithogalum
  mitans 78
  umbellatum 78
Orpine 138, 222
Orthilia secunda 71
Our Lady's Bedstraw 45
Ox-eye Daisy 93
Oxalis
  acetosella 65
  corniculata 142
  europaea 142
  stricta 142
Oxlip 149
Oxytropis
  campestris 332
  halleri 332
  montana 332

Pale Bugloss 340
Pale Persicaria 49, 214, 366
Pale St John's-wort 144
Pale Willowherb 210
Pale-flowered Orchid 178
Papaver
  dubium 208
  rhoeas 208
  sendtneri 36
Paris quadrifolia 371
Parnassia palustris 60
Parsley Piert 360
Pasque Flower 325
Pasqueflower
  Alpine 57
  Spring 85
  Yellow Alpine 86
Pastinaca sativa 147

ssp. *sativa* 147
ssp. *sylvestris* 147
ssp. *urens* 147
Pea
  Black *268*
  Narrow-leaved Everlasting *267*
  Sea *267*
  Spring *269, 333*
  Tuberous *268*
Peach-leaved Bellflower *319*
Pear *106*
Pedicularis
  foliosa 194
  oederi 194
  palustris 281
  sceptrum-carolinum 194
  sylvatica 281
  verticillata 280
Penny-cress
  Field *42*
  Perfoliate *42*
  Round-leaved *43, 292*
Peppermint *213*
Perennial Honesty *293*
Perennial Sow-thistle *173*
Perfoliate Honeysuckle *117*
Perfoliate Penny-cress *42*
Perforate St John's-wort *146*
Persian Clover *265*
Petasites
  albus 88
  hybridus 88, *243*
  paradoxus 243
Petty-leaved Spurge *128*
Petty Spurge *362*
Petty Whin *203*
Phacelia tanacetifolia 310
Pheasant's Eye *240*
Phyteuma
  betonicifolium 320
  nigrum 320
  orbiculare 321
  ovatum 320
  spicatum 77
Picris hieracioides 171
Pimpernel
  Blue *306*
  Scarlet *231, 306*
  Yellow *151*
Pimpinella
  major 69
  saxifraga 69
Pineappleweed *163*
Pinguicula

alpina *103*
grandiflora 347
lusitanica 347
vulgaris *103, 347*
Pink *304*
  Carthusian *216*
  Cheddar *217*
  Deptford *216*
  Maiden *217*
  Sea *232*
Plantago
  coronopus 45
  intermedia 302
  lanceolata 365
  major 365
  media 302, 365
Plantain
  Buck's-horn *45*
  Great *365*
  Narrow-leaved Water 32, 206
  Ribwort *365*
  Water- *206*
Platanthera
  bifolia 94
  chlorantha 94
Ploughman's-spikenard *161*
Polygala
  amara 336
  amarella 336
  chamaebuxus 205
  comosa 270
  serpyllifolia 336, 337
  vulgaris 270, 336
Polygonatum
  multiflorum 81
  odoratum 81
  verticillatum 81
Polygonum
  aviculare 49, 214
  bistorta 214
  hydropiper 206
  lapathifolium 49, 214, 366
  mite 206
  persicaria 49, 214, 366
Poppy
  Common *208*
  Long-headed 208
  White Alpine *36*
Pot Marigold *169*
Potamogeton
  natans 351
  polygonifolius 351
Potentilla
  alba 63

anglica 141
anserina 141
aurea 140
brauneana 140
crantzil 140
erecta 126
fragariastrum 63
palustris 223
reptans 141
sterilis 63
tormentilla 126
Prenanthes
  muralis 174
  purpurea 251
    ssp. *angustifolia* 251
Prickly Lettuce *176*
Prickly Sow-thistle *173*
Primrose *148*
Primula
  acaulis 148
  auricula 148
  clusiana 230
  elatior 149
  farinosa 230
  halleri 230
  hortensis 148
  x *pubescens* 148
  veris 149
  vulgaris 148
Procumbent Yellow Oxalis *142*
Prunella
  grandiflora 343
  vulgaris 343
Prunus
  avium 110
  x *fruticans* 111
  padus 111
  spinosa 111
Pulicaria dysenterica 161
Pulmonaria
  angustifolia 311
  mollis 311
  obscura 235, 311
  officinalis 235, 311
Pulsatilla
  alpina 57, 86
  vernalis 85
    var. *alpestris* 85
    var. *bidgostiana* 85
    var. *vernalis* 85
  vulgaris 325
Purging Buckthorn *379*
Purging Flax *65*
Purple Coltsfoot *243*

Purple Crocus 324
Purple Gentian 152
Purple Gromwell 313
Purple Loosestrife 241
Pyramidal Bugle 341
Pyramidal Orchid 253
Pyrenean Bastard Toadflax 48
Pyrola
  media 71
  minor 71
  rotundifolia 71
  secunda 71
  uniflora 70
Pyrus communis 106

Ragged-Robin 220
Ramischia secunda 71
Rampion
  Blue-spiked 320
  Dark 320
  Round-headed 321
  Spiked 77
Ramsons 80
Rannoch Rush 370
Ranunculus
  aconitifolius 57
  acris 134
  aquatilis 58
  arvensis 134
  auricomus 133
  bulbosus 135
  circinatus 58
  ficaria 158
  flammula 137
  fluitans 58
  lanuginosus 136
  lingua 137
  nemorosus 134, 136
  platanifolius 57
  repens 135
  reptans 137
  sardous 135
  sceleratus 136
  trichophyllus 58
Raphanus raphanistrum 38, 120
Raspberry 112
Red Bartsia 279
Red-berried Elder 201
Red Campion 219
Red Clover 264
Red Deadnettle 275
Red Hare's Lettuce 251
Red Helleborine 252
Redshank 49, 214, 366

Reseda
  lutea 126, 159
  luteola 126
Rhamnus
  catharticus 199, 378, 379
  saxatilis 199, 378
Rhinanthus
  alectorolophus 195
  angustifolius 195
  crista-galli 195
  minor 195
Rhododendron
  ferrugineum 290
  hirsutum 290
Ribbed Melilot 182
Ribbon-leaved Water Plaintain
  32, 206
Ribes
  alpinum 380
  uva-crispa 380
Ribwort Plantain 365
River Water-crowfoot 58
Robinia pseudoacacia 117
Rock Buckthorn 199, 378
Rosa
  arvensis 110
  canina 287
  dumetorum 287
  obtusifolia 287
Rosa gallica 287
Rose
  Christmas 56
  Dog 287
  Field 110
  Guelder- 116
  Snow 56
Rose family 287
Rosebay Willowherb 211
Rosemary Daphne 283
Rosemary Willowherb 211
Rough Hawkbit 171
Rough Hawk's-creed 175
Round-headed Leek 239
Round-headed Orchid 254
Round-headed Rampion 321
Round-leaved Bedstraw 46
Round-leaved Penny-cress 43,
  292
Round-leaved Sundew 59
Round-leaved Wintergreen 71
Rowan 109
Rubus
  caesius 112
  fruticosus 61, 112

idaeus 112
  saxatilis 61
Rumex
  acetosa 207
  acetosella 207
  alpinus 356
  conglomeratus 354
  crispus 355
  hydrolapathum 355
  maritimus 355
  obtusifolius 355
  sanguineus 354
  tenuifolius 207
Rusty Alpenrose 290

Sagittaria sagittifolia 32
Salad Burnet 361
Salicornia europaea 356
Salvia
  glutinosa 192
  pratensis 344
  verbenaca 344
  verticillata 344
Sambucus
  ebulus 76
  nigra 76, 115
    var. laciniata 115
  racemosa 201
Sanguisorba
  minor 361
  officinalis 210, 361
Sanicle 66
Sanicula europea 66
Saponaria
  ocymoides 221
  officinalis 55, 221
Sarothamnus scoparius
  204
Satureja
  acinos 345
  calamintha 345
Satyrium repens 95
Saw-wort 247
Saxifraga
  aizoon 61
  granulata 60
  paniculata 61
Saxifrage
  Alternate-leaved Golden 125
  Burnet- 69
  Greater Burnet- 69
  Livelong 61
  Meadow 60
  Opposite-leaved Golden 125

Scabiosa
  columbaria 348
  lucida 348
Scabious
  Devil's-bit 303
  Field 349
  Shiny 348
  Small 348
Scarlet Pimpernel 231, 306
Scented Mayweed 90
Scentless Feverfew 93
Scentless Mayweed 90
Scheuchzeria palustris 370
Scilla
  autumnalis 323
  bifolia 323
  verna 323
Scleranthus annuus 366
Scottish Asphodel 156
Scrophularia
  auriculata 377
  nodosa 377
  umbrosa 377
  vernalis 377
Scutellaria
  galericulata 342
  hastifolia 342
  minor 342
Sea Aster 326
Sea Bindweed 74
Sea Holly 306
Sea Lavender 307
Sea Mayweed 90
Sea Milkwort 231
Sea Pea 267
Sea Pink 232
Sea Rocket 37, 209
Seakale 37
Sedum
  acre 138
  sexangulare 138
  telephium 138, 222
    ssp. fabaria 138
    ssp. maximum 138
    ssp. ruprechtii 138
    ssp. telephium 138
Self-heal 343
Selinum carvifolia 69
Sempervivum tectorum 240
Senecio
  erucifolius 168
  helenitis 167
  jacobaea 168
  nemorensis 168

  ssp. fuchsii 168
  ssp. nemorensis 168
  spathulifolius 167
  sylvaticus 166
  vernalis 167
  viscosus 166
  vulgaris 166
Serrated Wintergreen 71
Serratula tinctoria 247
Shaggy Soldier 89
Sheep's Bit 321
Sheep's Sorrel 207
Shepherd's Cress 39
Shepherd's Purse 42
Sherardia arvensis 302
Shiny Scabious 348
Shrubby Milkwort 205
Sickle-leaved Hare's-ear 147
Sickle Medick 183, 332
Sickle Spurge 129
Silene
  acaulis 219
  alba 54
  cucubalus 55
  dichotoma 54
  dioica 219
  inflata 55
  noctiflorum 54
  nutans 54
  vulgaris 55
Silverweed 141
Sinapis arvensis 38, 120
Sisymbrium officinale 121
Skullcap 342
Slender Loosestrife 241
Slender St John's-wort 146
Sloe 111
Small Alison 122
Small Balsam 189
Small Cow-wheat 193
Small-flowered Evening
  primrose 130
Small-flowered Mallow 229
Small-flowered Winter-cress 120
Small-leaved Elm 285
Small-leaved Lime 201
Small Nettle 354
Small Scabious 348
Small Toadflax 347
Small Yellow Foxglove 196
Smooth Hawk's-beard 175
Smooth Honeywort 152
Smooth Sow-thistle 173
Smooth Tare 334

Sneezewort 92
Snow Rose 56
Snowball Tree 116
Snowdrop 83
Snowdrop Windflower 57
Snowy Mespil 107
Soapwort 55, 221
Soft Lungwort 311
Solanum
  dulcamara 315
  nigrum 75
Soldanella
  alpina 326
  minima 241
  pusilla 241
Solid-tubered Fumitory 97, 262
Solidago
  canadensis 160
  gigantea 160
  virgaurea 160
Solomon's-seal 81
Sonchus
  arvensis 173
  asper 173
  oleraceus 173
  palustris 173
Sorbus
  aria 108
  aucuparia 109
  torminalis 108
Sorrel
  Common 207
  Sheep's 207
  Wood 65
Sow-thistle
  Alpine 329
  Perennial 173
  Prickly 173
  Smooth 173
Sparganium
  angustifolium 118, 350
  emersum 118, 350
  erectum 118, 350
  minimum 118, 350
Spear-leaved Orache 358
Spear-leaved Skullcap 342
Spear Thistle 246
Spearmint 296
Spearwort
  Creeping 137
  Great 137
  Lesser 137
Speedwell
  Blue Water 297

Common Field 299
Fingered 299
Germander 301
Grey Field 298
Heath 300
Ivy-leaved 298
Thyme-leaved 44
Wall 300
Spergula
    arvensis 53
    petandra 53
Spiked Rampion 77
Spindle 378
Spiny Restharrow 291
Spotted Deadnettle 274
Spotted Gentian 152
Spreading Bellflower
    317
Spring Crocus 84, 324
Spring Gentian 309
Spring Pasqueflower 85
Spring Pea 269, 333
Spring Snowflake 83
Spring Squill 323
Spurge
    Broad-leaved 128
    Cypress 129, 362
    Dwarf 129
    Leafy 129
    Petty 362
    Petty-leaved 128
    Sickle 129
    Sun 128, 362
    Sweet 127, 362
    Wood 127
Square-stalked St John's-wort
    145
Squill
    Alpine 323
    Autumn 323
    Spring 323
St Bernard's Lily 79
St John's-wort
    Hairy 144
    Imperforate 145
    Pale 144
    Perforate 146
    Slender 146
    Square-stalked 145
    Trailing 143
Stachys
    alpina 276
    annua 101, 101
    palustris 277

recta 101
sylvatica 277
Star-of-Bethlehem 78
Stellaria
    graminea 50
    holostea 51
    longifolia 50
    media 50, 53
    nemorum 51
    palustris 51
Stemless Carline Thistle 91
Stemless Gentian 309
Stemless Thistle 247
Sticky Catchfly 220
Sticky Groundsel 166
Sticky Mouse-ear 52
Stinking Hellebore 367
Stitchwort 65
    Greater 51
    Lesser 50
    Marsh 51
    Wood 51
Stone Blackberry 61
Stone Bramble 61
Stratiotes aloides 33
Strawberry
    Barren 63
    Hautbois 62
    Wild 62
Streptopus amplexifolius 80
Striped Daphne 282
Succisa
    inflexa 303
    pratensis 303
Summer Snowflake 83
Sun Spurge 128, 362
Sundew
    Great 59
    Oblong-leaved 59
    Round-leaved 59
Sweet Flag 352
Sweet Spurge 127, 362
Sweet Violet 339
Sweet William 216
Sweet Woodruff 45
Swida sanguinea 105
Sycamore 116, 381
Symphytum officinale 153, 234,
    310

Tall Melilot 182
Tanacetum
    corymbosum 93
    vulgare 164

Tansy 164
Taraxacum officinale 172
Tassel Hyacinth 322
Tasteless Stonecrop 138
Tasteless Water-pepper 206
Teasel 303
Teesdalia nudicaulis 39
Tetragonolobus maritimus 187
Teucrium
    chamaedrys 272
    scordium 272
    scorodonia 191
Thale Cress 39
Thalictrum
    aquilegifolium 292
    flavum 118
Thesium
    montanum 48
    pyrenaicum 48
Thistle
    Alpine 245
    Cabbage 170
    Carline 169
    Creeping 328
    Marsh 173, 246
    Musk 245
    Spear 246
    Stemless 247
    Stemless Carline 91
Thlaspi
    arvense 42
    perfoliatum 42
    rotundifolium 292
Thread-leaved Water-crowfoot
    58
Three-leaved Valerian 76, 236
Three-nerved Sandwort 53
Thyme-leaved Speedwell 44
Thymus
    articus ssp. *praecox* 278
    serpyllum 278
Tilia
    cordata 201
    platyphyllos 201
Toadflax
    Alpine 346
    Common 197
    Ivy-leaved 346
    Pyrenean Bastard 48
    Small 347
Tofieldia
    calyculata 156
    ossifraga 156
    pusilla 156

Toothwort *281*
Tormentil *126*
Tormentilla erecta 126
Touch-me-not Balsam *189*
Tower Mustard 38, *121*
Tragopogon pratensis 172
   ssp. *minor* 172
   ssp. *orientalis* 172
   ssp. *pratensis* 172
Trailing St John's-wort *143*
Trailing Tormentil *141*
Trapa natans 43
Traunsteinera globosa 254
Traveller's Joy *104*
Treacle Mustard *121*
Trefoil 183
   Hop *185*
   Large Hop *184*
   Lesser *185*
Trientalis europaea 87
Trifid Bur-marigold *163*
Trifolium
   agraium 184
   arvense 98
   aureum 184
   campestre 185
   dubium 185
   filiforme 185
   hybridum 99, 264
   incarnatum 265
   medium 264
   minus 185
   montanum 99
   pratense 264
   procumbens 185
   repens 98
   resupinatum 265
   rubens 265
   strepens 184
   suaveolens 265
Tripteurospermum inodorum 90
Trollius europaeus 159
True Watercress *41*
Tuberous Pea *268*
Tuberus Valerian 237
Tufted Loosestrife *150*
Tufted Milkwort *270*
Tufted Vetch *335*
Tussilago farfara 88, *165*
Typha
   angustifolia 350
   latifolia 350

Ulex europaeus 202

Ulmus
   carpinifolia 285
   glabra 285
   minor 285
   procera 285
Umbellate Hawkweed group 177
Unbranched Bur-reed 118, *350*
Upland Enchanter's-nightshade 44
Upright Yellow Oxalis *142*
Urtica
   dioica 354
   urens 354
Utricularia
   australis 197
   vulgaris 197

Vaccinium
   myrtillus 289, 379
   oxycoccos 283
   uliginosum 113, 289, *379*
   vitis-idaea 113, 289
Valerian
   Common 237
   Marsh 237
   Mountain 236
   Three-leaved *76*, 236
   Tuberus 237
Valeriana
   dioica 237
   montana 236
   officinalis 237
   pratensis 237
   tripteris 76, 236
   tuberosa 237
Valerianella locusta 315
Vanilla Orchid *260*
Venus' Looking-glass *320*
Veratrum album 78, 152
   ssp. *album 78*
   ssp. *labelianum 78*
Verbascum
   blattaria 154
   densiflorum 155
   lychnitis 75, 154
   nigrum 154
   phlomoides 155
   thapsiforme 155
   thapsus 154
Verbena officinalis 235, *314*
Veronica
   acinifolia 299
   agrestis 298
   anagallis-aquatica 297

   arvensis 300
   austriaca 301
   beccabunga 297
   chamaedrys 301
   dillenii 300
   hederifolia 298
      ssp. *hederifolia* 298
      ssp. *sublobata* 298
   humifusa 44
   officinalis 300
   persica 299
   polita 298
   praecox 299
   serpyllifolia 44
   teucrium 301
   triphyllos 299
   urticifolia 301
   verna 300
Vervain 235, *314*
Vetch
   Bush *334*
   Common *266*
   Crown *263*
   Fine-leaved *335*
   Horseshoe *187*
   Kidney *184*
   Tufted *335*
   Wood *100*
   Wood Bitter 100
Viburnum
   lantana 115
   opulus 116
      var. *roseum 116*
Vicia
   angustifolia 266
   cracca 335
   dasycarpa 335
   hirsuta 334
   orobus 100
   sativa 266
   sepium 334
   sylvatica 100
   tenuifolia 335
   tenuissima 334
   tetrasperma 334
   villosa 335
Vinca
   major 307
   minor 307
Vincetoxicum *73*
Vincetoxicum
   hirundinaria 73
      var. *laxum 73*
   officinale 73

Viola
  biflora 190
  canina 338
    ssp. *canina* 338
    ssp. *montana* 338
  hirta 339
  odorata 339
  palustris 337
  reichenbachiana 338
  riviniana 338
  stagnina 337
  tricolor 190
    ssp. *curtisii* 190
    ssp. *subalpina* 190
    ssp. *tricolor* 190
Violet
  Bog 337
  Common Dog 338
  Dame's 293
  Early Dog 338
  Fen 337
  Hairy 339
  Heath Dog 338
  Sweet 339
  Water- 72
  Yellow Wood 190
Viper's Bugloss 340
Viscum album 198
  ssp. *abietis* 198
  ssp. *album* 198
  ssp. *austriacum* 198
Vogelia paniculata 123

Wall Germander 272
Wall Gipsy Weed 215
Wall Lettuce 174
Wall Speedwell 300
Wall-pepper 138
Wallflower 121
Water Avens 222, 368
Water Chickweed 51
Water-crowfoot
  Common 58
  Fan-leaved 58
  River 58
  Thread-leaved 58
Water Figwort 377
Water Forget-me-not 312
Water Germander 272
Water-lily
  Fringed 33
  Least 132
  White 84, 132
  Yellow 84, 132

Water Mint 213, 296
Water-pepper 206
Water Plaintain 32, 206
Water Soldier 33
Water-starwort 364
Water-violet 72
Watercress 41
Wavy Bitter-cress 40
Wayfaring-tree 115
Weld 126
Whin 202
White Alpine Poppy 36
White Bryony 77
White Butterbur 88
White Campion 54
White Cinquefoil 63
White Clover 98
White Deadnettle 100
White Dittany 271
White Helleborine 95
White Melilot 99
White Mullein 75, 154
White Pigweed 359
White Water-lily 84, 132
Whorled Clary 344
Whorled Lousewort 280
Whorled Solomon's-seal 81
Wild Angelica 69, 229
Wild Carrot 69
Wild Cherry 110
Wild Cotoneaster 286
Wild Liquorice 182
Wild Mignonette 126, 159
Wild Pansy 190
Wild Parsnip 147
Wild Privet 105
Wild Radish 38, 120
Wild Service-tree 108
Wild Strawberry 62
Willowherb
  Broad-leaved 213
  Great 212
  Hoary 212
  Pale 210
  Rosebay 211
  Rosemary 211
Winged Broom 181
Winter-cress 120
Wintergreen
  Chickweed 87
  Common 71
  One-flowered 70
  Round-leaved 71
  Serrated 71

Woad 123
Wolf's-bane 180
Wood Anemone 85, 87
Wood Avens 139
Wood Bedstraw 46
Wood Bitter Vetch 100
Wood Buttercup 134, 136
Wood Crane's-bill 227, 305
Wood Dock 354
Wood Forget-me-not 312, 313
Wood Ragwort 168
Wood Sage 191
Wood Sorrel 65
Wood Spurge 127
Wood Stitchwort 51
Wood Vetch 100
Woolly Buttercup 136
Woundwort
  Annual 101
  Hedge 277
  Limestone 276
  Marsh 277
  Yellow 101
Wych Elm 285

Yarrow 92, 244
Yellow Alpine Pasqueflower 86
Yellow Anemone 133
Yellow Archangel 192
Yellow Bellflower 155
Yellow Bird's-nest 130
Yellow Corydalis 181
Yellow Figwort 377
Yellow Iris 157
Yellow Loosestrife 150
Yellow Ox-eye 161, 162
Yellow Pheasant's-eye 158
Yellow Pimpernel 151
Yellow Star-of-Bethlehem 157
Yellow Vetchling 188
Yellow Water-lily 84, 132
Yellow Whitlow-grass 124
Yellow Wood Violet 190
Yellow Woundwort 101

Zigzag Clover 264